HUMAN ANCESTRY

WHO WE ARE; WHERE WE COME FROM

J NORMAN WILKINS

LifeRich Publishing is a registered trademark of The Reader's Digest Association, Inc.

LifeRich Publishing books may be ordered through booksellers or by contacting:

LifeRich Publishing
1663 Liberty Drive
Bloomington, IN 47403
www.liferichpublishing.com
844-686-9607

Because of the dynamic nature of the Internet, any web addresses or links contained in
this book may have changed since publication and may no longer be valid. The views
expressed in this work are solely those of the author and do not necessarily reflect the
views of the publisher, and the publisher hereby disclaims any responsibility for them.

Any people depicted in stock imagery provided by Getty Images are models,
and such images are being used for illustrative purposes only.
Certain stock imagery © Getty Images.

ISBN: 978-1-4897-3676-5 (sc)
ISBN: 978-1-4897-3675-8 (hc)
ISBN: 978-1-4897-3674-1 (e)

Library of Congress Control Number: 2021913026

Print information available on the last page.

LifeRich Publishing rev. date: 06/23/2021

CONTENTS

FOREWORD

Homo sapiens have a sense of awareness of self. Humans need to know who they are and where they come from. What is real must matter more than we wish to believe. Understanding the circumstances in the universe, origin, life, and the laws of science by natural selection are not spiritual quests. The spirit world principal of thinking and emotion for many individuals are from images of the non-living and living. The spirit world is not new. Humans have expressed their emotions of the unknown and practices of their belief and faith in the spirit world that have been passed on by generations through their ancestry. What are the differences and the lack of final destination about truth are the lessons in humility. This scientific historical study, with differences, supplements the ancestry of Homo sapiens with the knowledge of man's evolution, development, formation, and growth events. There are gaps in Homo sapien ancestry needing new fact findings and clarification.

Relation Theory by natural selection events, over billions of years provides clues about the non living origin of the universe, Earth, and the living life relationships beget (descendants) and begat (ancestors).

The origin from a non-living single state element in space created many universes when it expanded erupting in an explosion billions of years ago. Over time, our universe expanded forming the planet Earth as a plasma matter fire ball. Time passed, Earth had cooled forming a mantle of solid matter stratums.

The first Organic Micro-Molecule single cell life formed from mineral compound elements of Earth stratum by natural selection. Later, complex Multa-cellular life evolved that could reproduce and become larger in an Old World.

Micro life in the oceans produced larger forms of organisms. They became the creatures that could diverge, transform, and transition to new life forms. The larger and most fit creatures adapted to their environment and emerged unto the land as air breathing animals, too many genus to be classified. Over millions of years the Primates of two highly developed anamals; the gorilla-chimpanzee and monkey-ape diverged from each other in separate ways starting new separate primate lives in a New World.

In the New World our ancestors the Primate apes diverged as specialized variant apes. The fossils of Primates were discovered in Earth's stratum. Differences in bone structure, traits, and others clues linking them as ancestors; apeman, manape and man (Homo sapiens). Follow the clues about and understand nomad migration from Africa into Europe, Asia, and the Americas'.

With the emergence of Modern man, with the power of thinking and ability to perform tasks, formed social cultures, civilizations, and language established with a new way to communicate. Competition for control over others created aggressions that lead to violence and ill behavior, were punishable or may have included death.

After all men and some women claims of their status as chiefs of their tribes, some became leaders with power and control over others as authoritarians. Life, with its stress from emotions is responsible for adaptation from the environment and change by experience; knowledge, skill, and ability. Advanced computer data systems, artificial intelegence, robotics, and human life gene DNA engineering are the challenges evolving.

The Stone Age tools of Olduvai and Acheulean tool technology over time were used by hominids and Homo populations. Changing new technology worldwide opened up new opportunities and challenges for hominids and Homo.

The authors study of non-life origin, primate life, and the associated family links with its many gaps is a complex venture into ANCESTRY, of "Who we are and Where did we come from", events. Take some time to supplement some individual scientific clues about non-life, and life events. Please, read the text with a non-bias open mind, positive thinking, and be not judgemental or bias in reading these events. There is more to come as we continue to link and close the gaps in Homo sapiens family ancestry and the future from DNA research.

RELATION THEORY
NON-LIFE and LIFE MATRIX

N
      Natural Selection includes all of the events. Origin starts at the lower left of Non-Life
      World. Matter formed in four sequential events from left to right. Family Life starts
A
      at the lower right of the Old World progresses left from the bottom up. New World
      starts at the lower right and progresses left, from the bottom up.
T

      F
U   A   NEW WORLD
      M
R   I   HONO SAPIENS; (ANATOMIC MODERN, LATE MODERN, EARLY
      L   MODERN, MAN), MANAPE, APEMAN, SPECIALIZED GREAT APE
A   Y   AUSTRALOPITHECUS, COMMON APE.
               OLD WORLD
L

      L   MONKEYAPE – COMMON PRIMATE MAMMAL ANIMALS – WARM
            BLOOD VERTEBRATES ON LAND – ANCIENT CREATURES IN AN
            ALKALI OCEAN
      I
            MALE AND FEMALE DESCENDANTS EVOLVED;
      F   REPRODUCTION WAS ACCOMPLISHED WITHIN THE CELL
            NUCLEOTIDE – SPECIALIZED TRAIT TASK CELLS FORMED IN
            DUAL HELIX DEOXYRIBONUCLRIC
      E   ACID (DNA), CHROMOSOMES OF HEREDITY ARE GENETIC
            ALPHA CODED GENOMES MAPING EACH CELL STORED IN
            THE DNA BY NUMBER - METABOLISM; ACCOMPLISHED BY THE
            MITOCHONDRIA FROM WITHIN EACH CELL, PROCESSES CELL
            ENERGY BY ANABLISM, CATHOLISM, AND WASTE – AMINO ACID
            AND MOLECULE RIBONUCLEIC ACID ORGANIC PROTOPLASM
            COMBINE AS (RNA) - A MULTICELLAR CLUSTERED GROWTH
            CELL CONTAINS PROTOPLASM AND A NUCLEUS ENCLOSED IN A
            MEMBRANE – LIFE FORMED FROM ATOMS; CARBON, NITROGEN
            AND OXYGEN COMBINE TO FORM AMINO ACID.

            BACTERIA/ALGA – ONE CELL PHOTOSYNTHESIZED MICRO
            ORGANISM LIVED IN ANOXIC OCEANS DID NOT REPRODUCE-
            LIFE ORGANIC MICRO-MOLECULE FORM.

S   NON-LIFE WORLD
L
      M
E
      A
C
      T
T   GAS   PLASMA   LIQUID   SOLID
      T
I
      E
O
      R
N
O R I G I N   SINGLE STATE of EXISTANCE FROM A CENTRAL POINT IN
                   SPACE

INTRODUCTION

This ancient and modern technical scientific ancestorial relationship historical study is different in the presentation of events and ventures into conflicting dialog with Relation Theory by Natural Selection. Non-life origin events explore the universe of matter and the formation of Earth. Life events evolved as organic micro-molecule cellular development, micro-organisms, creatures, and for our purpose primate animals adapted to their environments of Earth's old and new worlds. Family origin is detailed by various forms of primate ape, apeman, manape, and man living and dead end events. The author technically eliminates some of the subjectivity, assumptions, opinions, and unknown distortions in the search for true and honest details using positive thinking behavior and testing standards to provide a greater understanding. Knowledge of changing times is the coming into existence of the evolution of Man's family valuations from clues, facts, theories, and anatomy of the primate animal body development over time to man. The survival of the most fit adapt to the environment. Basic and advanced knowledge from traditional education may be supplemented by these events.

Please, open your mind to non-living and living worlds presented within Origin Life and Family.

These events about evolution and human ancestry do not interfere with Pagan or Spirit World belief and faith.

The **non-living world** was a central point in space of a single state of existence, beginning from its explosion: Quarks were binding together, protons and neutron were formed among free electrons. The radiant energy of proton and electron particles in collision consumed the neutrons. Atomic nebula subatomic particles in the proton collision energy (condensed) slowed down those electron particles passing through to form the first atoms and are responsible for why matter has mass. The atomic nebula particles in collision emitted heat and light to begin. The first atoms were hydrogen and helium.

It is the unfolding of emissions of matter events that are responsible for the development of differences from their original to their present state. "**ORIGIN of MATTER**" is the coming together of elements of matter, mass, time, space, and energy by Natural Selection as a singular state of existence. From a central point in space a massive gas element matter fireball explosion disbursed gas and matter throughout many universes and everything within it. The origin of plasma matter formed the **Cosmo Universes** 14-13.7 billion years ago (bya). It can still be viewed by enhanced microwave observation today.

Over time the Cosmos was expanding and becoming larger. Development changed and included other denser matter by accretion. Some of its atomic energy elements expanded erupting into an explosion displacing a soup liquid matter evolving into many different formations.

There is an explanation for what is the origin of matter from the beginning of universes the first one billionth of a second of time. The cosmic atmosphere of gas, plasma matter, and atoms inflated into molecules of heavy element liquid matter and exploded into galaxies having gravity.

Over billions of years a **Solar System** atomic nebula particle of atoms condensed into a spinning proto planetary disk with a proto sun in the center and particle orbiting rings. Particle matter was drawn into the proto sun producing gas matter and exploded into a full plasma matter star, Earth and other planets sun. The energetic volatile gases from the sun fares were solar winds emitted into the outer particle rings. The particles grew larger by accretion. The larger particle bodies formed into planetesimals having gravity, established an orbit and spacing from the sun. The planetesimals became larger by accretion and some formed into planets.

Planet Earth was one of the liquid matter formations with no atmosphere. It was cooling with a solid matter mantle stratum (rock) and a plasma-liquid matter core. Earth was bombarded by meteorite impacts. Earth's moon was the collection of debris orbiting particles fused together by accretion and solar flares from Earth's sun and Earth orbit. Earth was no longer growing by accretion.

Earth's mantle cooled with the formation of solid matter (rock). Earth's violent volcanic liquid matter core activity mixed with other heavy atomic elements containing gases, liquids, and solids. The oceans filled with anoxic acid water and salts. Organic single cell micro-molecules formed the first life, but, could not reproduce.

Volcanic eruptions on the mantle from within the Earth's core activity formed new mantle solid matter. The surface of Earth was impacted by space debris, volcanic eruptions, seismic events, and ice ages glaciation. Time passed, the mantle became baron, ice subsided except for the north and south poles, oceans opened and closed, and the land continents continued to relocate geographically.

Time passed, organic multi-cellular micro-molecules that could reproduce became larger. These cells were the first to diverge into different forms of life. The most fit survived by eating each other, became larger, diverged as microorganisms, and diverged as ancient creatures.

Tectonic fracturing of Earth's mantle of convergent and divergent plate movements formed rifts and fissures destroying continents, ocean floor rock, causing earthquakes beneath the ocean, and tsunamis or tidal waves to occur devastating the land. Collision of land masses formed mountains. New continents and oceans were formed.

The volcanos continued erupting gas, lava, and ash liquid matter. The K-T asteroid event impacted Earth, turning the Earth ice cold. Fallout debris caused global fire over the surface, massive earthquakes, giant tsunamis radiated from the oceans, and most of the life became extinct 65 million years ago.

There are those who dismiss the origin of human life science as culturally and socially unacceptable. These prejudices were starting to change about the scientific acceptance of the evolution of Man by the 1860's.

Old World Pongidea Life evolved as many genera, species, and variants, too many to be classified. Organic micromolecule single cell life diverged as complex multi-molecule cells with the ability to grow larger into microorganisms, diverging and adapting to Ancient Creatures living and thriving by eating each other in an alkali changed anoxic ocean. Some grew into warm blood and cold blood Endothermics. Warm blood vertebrates evolved as Cynodonts on land and began to adapt to oxygen breathing air as well as living in the ocean. Primitive mammalian primates evolved as warm blood vertebrates diverging as animals having larger placental births. Over time there was a new order reorganization of the of primates; gorilla-chimpanzee and monkey-ape. They diverged and evolved as four separate species. Gorilla diverged from Chimpanzee and Monkey diverged from Ape.

Primitive creatures with mutant DNA markers, traits, and characteristics helped them to become adapted to their environment and tend to survive (survival of the fittest). They pass on their unique characteristics by reproduction and cell development performing different tasks for the next generation of growth by the process of Natural Selection. Living with stresses drives specie survival experiences; climate, food, and territory.

Old world Eomaia were the first placental mammals. They diverged, evolving much later as Purgatorius, the first mammalian primate-monkeys. The Plesiadapi-forms evolved as tree climbers monkey-like primitive primates. The Proconsul was a primitive mix form of advanced ape evolving and living in Africa. They were not ancestors of man. These apes were tree climbing quadrupeds. On the ground they used their hand knuckles to walk.

New World Placental Mammal Animals; the highest order evolved as Common Primates, diverging as Common apes forming their unique traits. They adapted to new environments, transforming as Great Apes. Dryopithecus fontani ape evolved in France, Spain, northern and central Europe and maybe east Africa. Sivapithecus ape evolved in East Asia and China diverging to Archaic Pithecanthropus ape. They were not hominids.

Homo sapiens ancestory is continually being recorded filling in the gaps from the Primate common ape divergence to the Super Family of Specalized Hominidae. "**ORIGIN of PRIMATE FAMILY**", over time New World variants diverged from the Great Apes. One of those was the Specialized Australopithecus (A.) ape becoming many species and variants.

Changes in gene mutant traits in body form evolved in transformed hominids and human species. Homo genus hominid ape was determined by brain size and did not diverge from the primate Australopithecus family. Homo species families transformed and transitioned as apeman, manape, and man.

The first major adaptive trait changes were a primitive transitional bipedal walking on the ground behavior in combination with primary quadruped palm walking life in the woodland trees by Ardipithecus Great Ape Chimpanzee variant. Primate A. anamansis shared some of the same traits at a later time. A. afarensis specie evolved with two major trait changes. The pelvis, muscles, and curved spine allowed for up right body balance for bipedal walking and allowed them to stride. The big toe was aligned with the other toes allowing for push off forward motion on one balanced leg. They would have been able to carry items in their arms or use tools to accomplish tasks.

Most of the Australopithecus lived in the East African Rift Valley System and others lived in South Africa. Apeman and manapes nomadic behavior allowed them to migrate out of Africa into Europe and Asia. They followed the animals of prey as scavengers of meat and gathered ground food to survive. Their calorie requirement was increasing as they ate more meat. Their adaptive traits changed the growth of the brain. This was the time when scientific authorities adopted the arbitrary concept of brain size. Australopithecus with brains larger than 600 ml were now classified as a new genus called Homo without having to diverge.

Homo (H.) habilis variants remained in Africa. H. ergaster variants migrated north out of Africa into Europe. H. erectus migrated into Asia and southern East Asia. They evolved, transformed and transitioned to apeman and manape H. sapiens migrated through out Africa, Europe, and the northern part of European-Asia. H. erectus vatiant beget Early H. sapiens living in South and North Africa. H. heidelbergensis controversity is the European preferred reference (skull cap) and may have been an early H. ergaster variant that beget H. antcessor (ANT). ANT beget H. neanderthalensis, Hono sapiens cousin.

Archaic Homo sapiens (AS) apeman transitioned to **Early Homo sapiens** (ES) manape, transitioned to Early Modern sapiens and Moderm Cro-magnon who were all man.

Early Modern H. sapiens (EMS) variants migrated through out Europe and Asia with major neurological trait changes in the brain related to thinking, and behavior development. At the same time early **Modern Cro-Magnon H. sapiens (CMS)** variants transitioned from ES with the same major neurological trait changes. **Late Modern CMS** migrated north into Europe from France. Both sapiens, all man, cousins were dominate species in Europe.

Late Modern sapiens (LMS) in Europe and Asia with innovation transitioned as the 1ˢᵗ European Asian sapiens (EAS). Their brains were developed with advanced thinking, cognitive traits, creativity, symbolic reasoning, and fully articulated speech. Many groups migrated from the EAS tribes.

European Asian sapiens (EAS) variants: The first group of EAS migrated into southern Siberia, Eastern Russia and Mongolia; part of that group divided and migrated into northern Siberia and become isolated in the mountains at the Artic Circle.

Another part of that group migrated east into Hot Springs, Beringia, Eastern Russia. They probably did not survive to migrate into the Americas'.

Another part of the first group of EAS migrated southeast over Beringia into Alaska Northwest Territory of North America. These EAS variants transitioned to the first Native American sapiens (FNAS), known as the First Nation People living in central southwest Alaska area. Some of that group migrated east over the Cordilleran Glacier to the Canadian Rocky Mountains Range at Alberta. The Canada passage way, south between the Cordilleran and Laurentide Glaciers, following the east flank on the eastern Rocky Mountain Range, and valleys leading to land in northern Montana, USA.

These LMS EAS FNAS variants were the first to be known as Eastern Paleo American. They transitioned to Native American sapiens (NAS). Part of the group migrated east along the ice sheet maximum to the Atlantic Ocean and transitioned to Modern NAS. Some of the group migrated south to Florida, around the Golf of Mexico into northern South America, and east to the Atlantic Ocean. Another group of the Modern NAS migrated throughout central USA and eastern Mexico. Others split at Panama and migrated along the Pacific Ocean western coast to the tip of South America.

Some of the original 1ˢᵗ group of LMS EAS migrated along the glacier maximium into eastern Asia in northern China and southern Beringia, Eastern Russia, at the Pacific Ocean. They became isolated by the ice blockage near the Pacific Ocean for 10,000 years. Some of the LMS EAS group in China migrated south along the Pacific Ocean into Korea, Indochina, Malaysia, and Indonesia; others migrated over land bridges when sea levels were -400 feet to the Nusa Tenggara Islands, Sulawesi Islands, and Maluku.

The LMS EAS from the 1ˢᵗ group at the Pacific Ocean, were isolated for 1000 years, beget with H. Pithecanthropus and transitioned as European

East Asian/East PI sapiens (EEAS) variants in northeast China. When the glacier ice started to open 20 tya some of the EEAS group migrated north crossing the land and ice bridges over Beringia to the Alaska Northwest Territory along the Cordilleran Glacier and the Pacific Ocean to Washington State, USA, on the west side of the Rocky Mountains Range as Paleo Americans. Part of the group of EEAS migrated down the western Pacific coast to the islands off the coast of what is Los Angles, USA.

These EAS and EEAS variants were unique from different areas, names, languages, and social cultures. The cultural and social diverse complex civilizations had developed from a verity of adaptations to the environment and technology.

Anatomic Modern H. sapiens (AMS) transformed from LMS as all man, with a neurologically changed frontal brain development. Some AMS migrated out of Africa into Europe and Asia. AMS became the dominate power in Europe. They had no tolerance for ill behavior and exiled others into isolation where it was difficult to survive, notably the H. neanderthalnesis.

More violence and ill behavior developed by the male species with high testosterone hormone characteristics. They were lacking the education needed to control their emotional stress. There was pilfering (taking what was not theirs from those who have) bloodlust with vengeance, and the emergence of authoritarian rule. Change and education was needed to surpress physical violence, mental health, and crime.

Social culture and civilization started when population increased. **Homo sapiens** changed with social differences, appearances of class, and categories. Authoritarian rules by the non productive population control the uneducated population of servants who produced the goods, bought the land, and paid taxes as land holders. Power over populations evolved in greed, violence, and warfare. Sounds like the dark sges in Europe's Vikings and Teutons are with us in todays world.

Modern technology demands educated populations with high technical skills in a computerized data world, using artificial intelligence, and robots performing industrial and other tasks. We pass on our changes in DNA traits from our experiences to the next generation as good, bad, and indifferent ill deeds in our genes. Leaders and followers must be responsible and accountable for their actions in earning the trust of the population.

PART 1

RELATION THEORY
NATURAL SELECTION

Origin of Matter Non-Life

The **non-living world** from a central point in space as a single state of existence, was the beginning when matter, time, space and energy all come together. **Existence** is a thing of any matter and energy that exists. The thing was inanimate. Hot and dense gas matter energy potential force capable of action was expanding. It exploded in a massive fireball some 14-13.6 billion years ago (bya) forming many new small expanding universes. Cosmic Microwave Background Radiation (CMBR) is a computerized focused and enhanced technology that can detect the energy image of the universes. It was the only light source we have discovered that still transmits this event and can be visually viewed as the faint heat radiation point in space that emits uniformly to all universes.

NUCLEOSYNTHESIS EVENT

"Creation" by Neucleosynthesis was the event of coming into existence of the universes and everything in it. In the first Femtosecond (10 to the -15 power = p) through a Picoseconds (10 to the -12 p) of one second our universe was a soup of scores of particles that varied enormously. It was followed by ¾ of a Nanosecond (10 to the -9 p) of one second or one billionth of a second through one microsecond (10 to the -6 p) of one second and ¾ of a millisecond (10 to the -3 p) of one second, Quirk energy was binding electrons from space atmosphere into two different types of quirks (q) to combine; a proton was made of two up q and one down q, a neutron was made of two down q and one up q, and the free electrons were without collision.

The size of our new universe is now hundreds of millions of miles in diameter and the temperature in the tens of trillions of Kelvins (K) degrees above absolute zero -460 Fahrenheit (F) or -273 Celsius (C). One second to 180 seconds after the explosion the protons were destined to become the nuclei of hydrogen atoms. One second to one minute (10 to the 0 p) into the new universe Nucleosynthesis collisions released energy (10 to the 18 p) between protons and neutrons forming new nuclei. One hour (10 to the 8 p) other light weight atomic energy elements formed as hydrogen and helium (10 to the 20 p), and at one day lithium and beryllium elements were formed. Over a three minute period of time the process formed the nuclei of 98% of the helium atoms present in space atmosphere today. That energy absorbed all of the neutrons.

Atoms Event

The universe continued to expand and cool for 100's of thousands of years. It was too energetic to form atoms. In the cosmic atmosphere of hydrogen and helium gas matter the atoms produced energy of heavier more dense helium reducing the atoms to carbon from the intense temperature, ionized the atoms stripping their electrons in plasma matter, and consumed other heavy elements into more carbon. These larger matter concentrates were influenced by other matter producing **molecules**; the smallest particle of an element or compound that can exist in a free state and still retain the characteristics of the elements of energy that composed the **atoms**. Those micro particles consisting of electrons revolving around a positively charged nucleus, containing protons and neutrons combined, to make a molecule.

When the electrons and atomic nuclei came together momentarily they were quickly split apart by the photons trapped in the continual collision with particles. The temperature had cooled to 3000 Kalvin and time had pasted 300,000 years. At the same time swirling clouds of photons, quantums of light energy were released streaming freely in all directions. At this stage the universe became transparent, as a fog of earlier particles and energy cleared.

Subatomic particles contained in the colliding protons produced heat and light that were the micro building blocks passed through the particle, stucking together, slowed down the energy. This process started the protons and atomic nuclei to capture electrons permanently forming the first atoms. These subatomic particles are responsible for why matter has mass found in colliding protons, primarily the formation of the elements of hydrogen and helium gas atoms.

Galaxies Event

Gas atoms were coming together because of gravity. Over 100 thousand million years swirling clouds of hydrogen and helium formed in long thin strands. Thirteen bya or 500 million years (mya) after the beginning of **Gas Matter** explosion strands begin to clump together to form the first galaxies. Extreme high pressure and temperature increased and further energy concentration of matter within the galaxies led to the evolution of the first stars. The stars energy produced heat and light when the hydrogen nuclei center starts fusing to form helium. New heavier elements evolved from lighter ones. The very heaviest elements only form in large star supernovas, until there was an explosion of the cosmic matter, distributing new **Liquid Matter** through out the galaxies incorporated in new stars and planets. These events expended the matter from the explosion containing carbon, iron with magnetic and electronic fields, into other heavier liquid matter that no longer supported the helium environment. The matter condensed, forming a new liquid matter made by combining of those particles subjected to specific gravity. Cassiopeia A supernova exploded 10 bya. The Milky Way was the first to form approximately 11-10 bya.

Solar System Event

The formation of gas and dust nebula began to condense, shrinking into spinning proto-planetary disks, with a protosun center and rings about 4.5 bya. The nebula matter coalesced into a dense center region as a proto sun, with diffused outer regions. Over time shrinking proto-planetary disks of ice and dust collided to form large particles. In the center of a large disk gravity drew in ice and dust partical matter, the temperature rose, hydrogen fused to form helium and the disk exploded into a full star, and our sun of **Plasma Matter** took its place in our universe. The sun is denser than the element of lead, having a core temperature of 15 million degrees centigrade. The sun radiates heat, light, and stream of energetic volatile solar flares, known as the solar winds, from the surface outward into the plantesimal disks. As particles in the disk became larger gravity drew the particles into collisions by **Accretion**. Larger masses formed, composed of rock or rock and ice. Planetesimals established an orbit and spacing depending on the internal gravity influence of its iron composition matter. Not all planetesimals became planets, but, became asteroids and comets. Asteroids orbit between Mars and Jupiter and cross paths with Earth's orbit. Comets formed from icy planetesimals in the outer edge of the disk. Most of the rock planets orbited within the four inner disks.

Earth System Events

Planet Earth was formed in a geological time of cataclysmic change. The matter transformations of thermo-electromagnetic accretion formed our Earth. The hot liquid matter planet had no mantle or atmosphere 4.56 bya (4560 mya). The bombardment of planetesimals, the differentiation of the core coalescing rock debris, orbited Earth as it cooled. Earth's violent volcanic activity mixed with other heavy element matter thickened into a soup like mixture, and was further inflated into other matter. This matter cooled as **Solid Matter** into layers that developed Earth's mantle. Earth no longer grew by accretion. From the early accretion of coalescing rock debris orbiting Earth, the satellite accretion became Earth's moon.

The evolving ecosystem was changing. Volcanism, meteorite bombardment impact events, and climate changes culminated in runaway glaciations. The first oceans floor rock was forming and being destroyed. The first organic molecules formed in the ocean 4.2 bya. Continents were growing through the process of plate tectonic margins of convergence and divergence having an effect on evolution and life. Earth's first primitive atmosphere was bombarded by meteorites and solar winds vaporizing the oceans releasing hydrogen and helium into space by 3.8 bya.

The solid matter layers continued forming Earth's mantle. Low density silicate minerals accumulated, forming Earth's outer mantle. Volcanic eruptions continued, resulting in the formation of new surface rock stratum. Volcanoes produced acid and condensed water vapor forming anoxic oceans. The salts slowly dissolved and increased salinity about 3.75 bya.

Snowball Glaciation Event

The Snowball Glaciation Event of the ice caps extended from Earth's poles into the tropics. Volcanism caused climate change culminating in glaciations 2.6 (2.45-2.22) bya, and again 800 mya.

The cooler and denser rock formed earlier sank into the interior. The mantle rocks fragmented into plate tectonics with both convergent and divergent margins. Internal low density rocks accreted into continents and higher density rock formed the ocean floor 1.8 bya. The continents were clustering to form Rodinian super-continent 1.35 bya. The Earth's tectonic plates had a tremendous influence on evolution of life. The first birds Archaeopteryx evolved 150 mya. The first flowering plants, feathered dinosaurs and placental mammals evolved 124 mya.

Time Events

Time Values are expressed as billions of years (bya), millions of years (mya), and thousands of years (tya). 9000 tya is 7000 BC using 2000 AD as a base.

- PRECAMBIAN EON 4.56 bya-543 mya Eon is a very long periods of time.
- HADEAN ERA 4.56-3.8 bya Era is a period of time in number of years from a given date.
- ARCHEAN ERA 3.8-2.5 bya
- PROTEROZOIC ERA 2.5 bya-543 mya
- PHANEROZOIC EON 543 mya to present time
- PALEOZOIC ERA 543-252 mya
- CAMBRIAN PERIOD 543-490 mya A period is the interval between the successive occurrence of astronomical events.
- ORDOVICIAN PERIOD 490-443 mya

- SILURIAN PERIOD 443-418 mya
- DEVONIAN PERIOD 418-354 mya
- CARBONIFEROUS PERIOD 354-290 mya
- MISSISSIPPIAN EPOCH 354-323 mya Epoch is the period of time considered as a noteworthy event historically.
- PENNSYLVIAN EPOCH 323-290 mya
- PERMIAN PERIOD 290-252 mya
- MESOZOIC ERA 252-65 mya
- TRIASSIC PERIOD 252-199.5 mya
- JURASSIC PERIOD 199.5-142 mya
- CRETACEOUS PERIOD 142-65 mya
- LOWER CRETACEOUS EPOCH 142-99 mya
- UPPER CRETACEOUS EPOCH 99-65 mya
- CENOZOIC ERA 65 mya to present time
- TERTIARY SUB ERA: PALEOGENE PERIOD 65-23.3 mya
- PALEOCENE EPOCH 65-54.8 mya
- TERTIARY SUB ERA: NEOGENE PERIOD 23.3-1.8 mya
- EOCENE EPOCH 54.8-33.5 mya
- OLIGOCENE EPOCH 33.5-24 mya
- MIOCENE EPOCH 24-5 mya
- PLIOCENE EPOCH 5-1.8 mya
- QUATERNARY PERIOD 1.8 mya to present time
- PLEISTOCENE EPOCH 1.8 mya to 10 tya
- HOLOCENE EPOCH 10 tya to present time

Paleo-Magnetic Event

Earth's magnetic force field extends from pole to pole. The magnetic characteristics of tiny grains of magnetic material in the iron element recorded the direction of Earth's magnetic field at the time of its origin. Earth's magnetic poles reversed 780 mya during the late Proterozoic Era and can last hundreds of thousands of years. What was magnetically south is now north. All samples of rock and stratum after the Era point north and those before point south. There were banded of iron formations in the oceans by 750 mya. Pale magnetic data combined with Electron Spin Resonance testing provides time values with some errors in tested samples.

The testing process of stratum or rock where a fossil was discovered is used to determine the approximate age of the fossil. Single Crystal Laser Fusion dating technology; uses the sample of rock sealed in glass as it was extracted on location, marked with the North-South orientation for geomagnetic shift which occurs at known intervals over time, and a hand held compass device is placed in the laser fusion chamber. The laser energy melts the potassium feldspar crystal, which produces the resulting argon gas, measured in the gas mass spectrometer for radioactive decay of the potassium, revealing the sample rock age with an error analysis of less than 1%.

Earth's magnetic polarity has reversed over time in rocks containing iron. Some expert formed opinion that Polar reversal may have been from glacial sediment from all over the Earth. Other opinion statements of the reversal may have happened when our solar system Sun Plasma Flare Corona (solar wind) magnetically charged positive were attracted to Earth's iron element grain deposits, disrupting and reversing polarity or magnetically oriented elements in rocks with iron content. Plate tectonic; divergence, convergence, and continent drift movements, violent volcanic activity, earthquakes, ocean floor iron deposits movement, and organic minerals in sediment locked in glacial formations all influenced change in rocks with iron content, from magnetic South Pole to magnetic North Pole. This occurred with many retreats in between. Magnetic rocks containing iron rich grains clustered near the South Pole were moved by plate tectonics to North Africa. At the same time, the rotation of Gondwanaland was rifting and moving toward the South Pole.

Tectonic Events

The Continental Plate tectronic movements caused land and ocean changes in different climate zones 600-250 mya. Tectonic; divergence, convergence, continent drift movements, violent volcanic activity, earthquakes, ocean floor movement, and organic minerals in sediment locked in glacial formations all influenced change.

Most of the continents clustered in the southern hemisphere. Tectronic clustering continued and formed Gondwanaland super-continent 600 mya. Over time it drifted north from the southern hemisphere to form an even larger super-continent called Pangaea by 550 mya. It eventually stretched from pole to pole. North America rifts from Gondwanaland, which was moving south rifted by the Iapetus Ocean 530 mya. Laurentia and Baltica were converging 440-290 mya. The shallow seas were flooded by the highest sea levels. Avalonia moves north as the Iapetus Ocean closes by 420 mya. The land was desert 418-354 mya. The above three amalgamations formed Laurussia 410 mya. The shallow seas were flooded by the highest sea levels.

The seas were tropical 354-323 mya. There were tropical forests 323-290 mya. Gondwanaland moved north causing sea floors spreading, resulting in the South Pole glaciations 320 mya. The land was again a desert 290-250 mya. Those changes were caused by the continental plate tectonic movements into different climate zones.

The volcanism associated with plumbs of heat from within the Earth's core led to doming and rifting. Vast outpours of lava and volcanic gasses from the Siberian Traps altered the global climate 252 mya. There were widespread anoxic oceans and sea levels below 450 feet 245 mya. The sea floors were spreading and were continuously being reformed 235-110 mya. Volcanism, surface doming, rifting, and the Pangaea breakup of rifting caused the sea levels to increase 235 mya. There was an opening between Gondwanaland and Baltica forming the Tethys Ocean 220 mya. North Africa, Europe, and Asia separated with the rapid fluctuating climate as another ice age started 186-170 mya. The Golf of Mexico and mid Atlantic Ocean opened up 170 mya. Southern Atlantic Ocean starts to open 153 mya. Madagascar rifts from Africa 151 mya. India rifts from Antartica 138 mya and moves rapidly to the north by 82 mya. The Tethys Ocean begins to close and the diverging plate rift was building the Alps. The India Plate was converging with the diverging Asia Plate and began to build the Himalayas 90-20 mya. The North Atlantic was opening 85 mya. Laurussia breakup event was 75 mya. In North America, rifting formed the western flank of the Rocky Mountains in Alaska to New Mexico, USA 70 mya.

The East African Great Rift System tectonics started 60 mya. The Arabian Peninsula Plate diverged forming the Red Sea depression from Cairo Egypt to the Gulf of Aden. The Nubian Plate northwestern movement opened at the south end of the Red Sea Awash Valley System western margin. The Somalian Plate southeast movement opened the eastern margin at the southwest end of the Red Sea, at Eritrea, at the eastern margin and the mouth of the Golf of Aden, at Djibouti, Somalia, forming the Awash Valley Rift System, Afar Basin, and the Olduvai Gorge. The rift runs through Ethiopia, Kenya and Tanzania. The formations cracked into slabs and were strewed about haphazardly 32-25 mya.

The Ethiopian flood of basalt lava outpour occurred 31 through 28 mya. Near the village of Herto, Ethiopia, the volcanic flow formed the Bouri Peninsula blocking the west and east margins of the Olduvai Gorge, forming a volcanic natural dam. It forms the north boundary of Yardi Lake and the mouth of the Awash River flowing northwest to the Red Sea through the Awash Valley System, Afar Basin.

Eruption of Sadiman volcano 5.2 mya deposited lava and thin seams of tuffs interlaced among the ash over the enormous Serengeti flood plains in Africa. Over time sediment built up on top of the basalt base and tuffs. Erosion slowly changed the landscape and environment. Located on a ridge the seams of volcanic ash tuffs of silica glass, Lusaka Tuft, "lion hair", in the Afar language, are not radio-metrically datable due to the lack of minerals.

In North America there was an outpour of lava in the Columbia River area in Oregon, USA 15 mya.

Atmosphere Event

There was no ozone outer layer atmosphere protection for Earth 4.15 bya.

Volcanism built up the second atmosphere releasing nitrogen, Carbon dioxide (CO_2), and water vapor. The water vapor was split by Ultra-Violet (UV) light into hydrogen, Oxygen (O_2), and ozone. The lightest gas hydrogen was released into space.

Organic micro molecules were primitive life photosynthesized from the sun in the oceans, releasing O_2 into the early atmosphere 2.7 bya. Most of the O_2 was used as an oxidant of Earths iron formations. Only 1% escaped into the atmosphere. The atmosphere O_2 level was 15% by 2.2 bya. Only 2% of O_2 formation was used as a protective ozone layer above the atmosphere acting as a filter of UV light by 1.9 bya. The O_2 level was 18% and the temperature was high by 5.4 mya.

The CO_2 was forming at a rate of 16x (when x is normal of the then present level) greater than the present levels by 500 mya. The temperature was decreasing in cool marine waters by 490 mya. By 460-450 mya the temperature was below average (59.9 For 15 C) and the CO_2 was 17x of normal by 450 mya. CO_2 was 12x and O_2 was 15% by 380 mya.

The release of large quantities of CO_2 gas from frozen masses of Methane organic compound (CH-4) a poisonous gas to most living organisms was produced from decomposed vegetation traped in glacier ice 260 mya. CH-4 gas hydrates lain buried in ice scabbed deposits were released with the melting and breakup (caving) of large sheets of ice and icebergs into the oceans. It disrupted the food chain, causing 90% extinction of all living organisms, and 60% of all genera both marine and land living organisms became extinct by 252 mya.

The CO_2 was decreasing and O_2 was increasing to 19% and then fell 230 mya. Between 190 and 50 mya every quarter, O_2 was at 15% and slowly increasing to 22%, 24% and 27%, CO_2 level was 3.5x and subsequently increased to 4.5x, then decreased to 3x, 4x, and 2x. The average temperature was 66F/19C, 63F/17.5C, 61F/16C, and increased to 63F/17.5C.

The K-T layer represents debris that was suspended in the atmosphere blocking the sun's rays, turning Earth ice cold with glaciation at the poles 67-64 mya.

Climate Event

An Ice Age started 445 mya. The temperature was increasing by 440 mya. The Siberian Trap volcanic eruption outpours of lava and gasses altered the global climate 252 mya. North Africa, Europe, and Asia separated with the rapid fluctuating climate as another ice age started 186-170 mya. Earth was cold with large areas of ice formations. The climate became increasingly arid as the Earth slipped into an Ice Age 4.5-4.3 mya.

Foraminifera's are shell compositions of tiny sea dwelling single cell crustations recording changes of ocean water chemistry. From the recovered fossils from ocean floor sediment, analyzing their chemistry over successive generations we can reconstruct the changes in ocean chemistry. Oxygen isotope ratios in the calcium carbonate of the shells are measured and from this information past fluctuations in ocean temperature, ice volume, and climate change can be recovered.

The sediment revealed a long term shift to a more arid climate. Driving these changes was the major cooling and drying. Early cyclic climates 3 mya ended cold (C) 900 tya. Another ice age began and climate in the high latitudes in Africa was cooling by 2.6-2 mya. It forced the hominids to migrate to new hard to find food. The hominids abandoned the forest and moved onto the savanna. There was competition for resources in a changing environment.

South Africa was drying 2.5 mya. The forests were turning into grasslands as the animals migrated adapting to their environment or became extinct. The drought continues in Central, East and North Africa.

The southern Great Rift Systems margins were fringed with woodland and savanna. The Olduvai Gorge was a salt lake 2.6 mya and fresh water 1.2 mya-620 tya. Tropical forests were being confined to central and west Africa. The Sahara became a vast region of savannah with lakes, rivers and woodlands extending eastward into Saudi Arabia, India, and China. The sea levels were low as the polar ice caps and ice sheets grew over Europe, Asia, and North America.

During the Pleistocene epoch 1.8 mya-10 tya the climate varied from warm (W) to C. It was C 1.8 mya to W 900 tya. The cyclic environmental changes had important influence on survival of all species. The adaptation and developmental changes evolving from those events, over time, established coping tolerances for the humans.

The eruption of the Deccan Traps basalt lava and ash flow caused global climate changes in conjunction with K-T asteroid impact events were responsible for 50% of all species death. This caused the Earth to turn very cold with glaciation at the poles 67-64 mya. Changes in the climate continued to open the North Atlantic Ocean 60 mya. The layer represents debris that was suspended in the atmosphere blocking the sun's rays, turning Earth ice cold with glaciation at the poles. K-T was responsible for 50% of all species death 67-64 mya.

The cooling climate in Africa caused deserts to form and forests disappeared changing to grassland 16 mya.

The Artie ice sheets begin to form locking up increasing amounts of water and snow 10 mya.

The climate became increasingly arid and the Earth slipped into an ice age 4.5-4.3 mya.

There was another Ice Age begining in the northern latitudes 900 tya.

It was C and wet 800 tya and cyclic CWC to 700 tya. The middle climate started 700 tya and deteriorated by 650 tya in Britain, United Kingdom. Humans and animals retreated across the dry land connecting the British Isles to the continent at that time. The climate was W 600-100 tya and humid. The north shore of the English Channel, in the British Isles, southeastern Britain's land bridges at 7 mile inlands was the shoreline 480 tya.

The climate 600 tya was cylic WCWCWC to 500 tya and very cold in Asia. During the Anglian glaciation 470 tya it became colder and the humans retreated south. It was another 250-200 thousand years before humans returned to Britain and northern Europe. There was a brief time period W 400-380 tya during the Hornian interglacial period when the temperature was 40-55 F and rainfall was around 32 inches.

By 350 tya glaciers covered northern Europe and Asia. Africa was starting to become arid C 200 tya and was cyclic WCWCWCW to 186 tya. Another Ice Age began 186-170 tya. It sealed off 17 million square miles of Earth and the sea level dropped 400 feet. It covered most of Europe, Asia, and North America.

The maximum glaciation was an irregular glacier line over Europe from Swanscombe, United Kingdom, Biachest Vaast France and southern France, Spy Belgium, Neander Germany, East Mladec Czech Republic, East Uzbekistan, Russia (European, Central, and Eastern), North America (Canada and USA), South Israel, eastern Spain, Italy, Croatia, and Hungary lasting until 170 tya. The humans migrated and abandoned the glacial maximum to the southern regions, ending C 120 tya.

The climate was cooler and dryer caused by North Atlantic iceburgs cooling the ocean 90 tya, the climate was cylic WC to 70 tya. Another Ice Age began 70 tya ending C 10 tya. The seas decreased to 400 feet below sea level, as the water was absorbed in glaciers. The Ice Age again reaches the Iberian Semiard Steppe 30-23 tya and blocked further migration into Europe, Asia and North America 30-18 tya. Global temperature decreased again, at the lowest point 23 tya when the Pennines Ice Sheets advanced from the north during the peak of the ice age, covering most of Europe except the southern region, southern Asia and southern North America 22-20 tya. The humans again, migrated and abandoned the glacial maximum to the southern regions. North America and the Northern part of Unites States were covered by the Cordilleran and Laurentide Glacier Ice Sheets and were in the process of deglaciation 21-10 tya. It was C 10 tya and ended W 9 tya.

K-T Asteroid Event

The asteroid event occurred at the end of the Cretaceous Period about 65 mya. An asteroid that may have been 6 miles wide impacted the Earth centered on the town of Chiexulub, Mexico, now beneath the Yucatan Peninsula, was over 185 miles in circumference. The asteroids leading edge collapse from the impact as the back continued forward. A crater was formed 60 miles wide and 7.5 miles deep. The crater collapsed inward for 150 miles, as the steep sides fell in burning debris 2/3rds of a mile beneath the Peninsula.

The rocks were blasted into the atmosphere. Upon reentry the rocks radiant heat caused global fires that raged over Earth's surface creating massive earthquakes, giant tsunamis radiating across the oceans, creating giant tidal waves at speeds up to 100 miles an hour, devastating the land, oceans, and living populations. The thin layer of clay dated from the same time period was sandwiched between the layers of Cretaceous Period (K) and Tertiary (T) sub Era in the paleogene carbonate rocks creating what was called the K-T boundary. This clay had high concentrations of iridium elements found in abundance in meteorites. The layer represents debris that was suspended in the atmosphere blocking the sun's rays, turning Earth ice cold with glaciation at the poles. K-T was responsible for 50% of all species death 67-64 mya.

Origin of Life

Non-life events happened for life to evolve. The origin of life process is the first primary focus. **Form** is of the body or figure of an animal or person. **Divergence** is the variations from the normal; difference; deviation that turns aside from a standard. **Transformed** is a change in the form or appearance from breeding (beget), Example: H. habilis to H. erectus, H. erectus to H. sapiens. **Transition** is the passing from one condition to another condition or anything essential to the existence or occurrence of something else, Example: ape, apeman, manape, man, and the physical changes in body anatomy form with features similar or same as man. **Homo** is a genus of primate and includes man. **Genus** is the main subdivision of a primate family and includes one or more species and variants. The genus is capitalized followed by the species not capitalized. **Species** have the appearance, shape, and distinct kind of animal having distinguishing characteristics of the genus. **Kind** is a natural group or division. **Variant** is the degree of change or difference in any way from others of the same kind. **Primate** is any member of the most highly developed order of animals composed of apes and the person man. **Man** is a **human;** having the form or nature of a person. A **person** is an individual human, distinguished from a lower animal, male or female. **Animals** are any living organism typically capable of moving about, <u>not making its own food</u> by photosynthesis. **Common** for our purpose is shared by belonging equally to all primates. **Pagan:** person having no religion.

The atmosphere and climate played a role in primate evolution and adapting in key events when the apes shift from the trees to the ground, progressively transforming and transitioning, with increasing reliance on brain power to survive.

Organic Micro-molecule life formed from sedimentary hydrocarbon stratum (rock) forming a new chemical compounds of acids and salts. **"ORIGIN of LIFE"**, are the genera, species, and their variants evolving over generations by **Natural Selection** in Earth's old and new worlds. Living with characteristics that help to become adapted to the environment, tend to be the survivers of the fittest, transmitting unique characteristics to the next generations. These characteristics are the distinctive distinguishing traits or features of the original makeup form and the appearance of the primate animals.

The Earth's tectonic plates had a tremendous influence on evolution of life. **Evolution** is the unfolding process of development, formation, or growth from earlier forms, having evolved from ancestors by heredity (**Begat**). **Beget** is the cause to exist, evolving from a later form by breeding, the transmission of slight gene variations in successive generations. Life **Existence** is the actual living fact of a **Being**; a thing of any matter and energy that **Exists**; life that is capable of sensation and motion shared with **Organisms**; of any living thing, capable of sensation and motion in its complexity of its composition, especially quadruped, but it excludes man. Man evolved as a primate animal.

Living humans closely related share a common ancestor. Variation in primate bone in numerous fossil records defines new genus, species, and variants. DNA differences in genetic makeup between two species, not mixed by breeding, is caused by their mutated genes built up in both species genomes becoming unique to only their specie.

DATING IDENTIFICATION TECHNIQUES

Many types of digital and computerized technology tools are used in finding the facts about human ancestory. Test the discovery that passes the test. Reject the test that fails. Follow the evidence where ever it leads, question everything, including the authority.

Radiation ray exposure is used in many testing methods. This testing device measures the radiation exposure values as ray radiation exposure in millisievents (m). The values are averages sized material and have differences in imaging. Natural Radon (r) exposure is present in the materials discovered exploring the environment we live in. The range of radiation used in various devices; 0.005m compares to 1r day exposure, Other devices use radiation exposures as high as 25m compared to 8r years.

Relative Dating is a visual analysis determination from the surrounding rock or **stratum**; a horizontal layer of matter or several layers of sedimentary, rock, and other animals that had coexisted during Hominid primate life helped in identifiable evolutionary adaptation within specific frames of time. Archaic fossils were not dated. Age analysis was determined through biochronology and paleomagnetism.

The **Molecular Clock** concept of time events are estimated times determined by molecular clock. Measuring the genetic similarity between living groups, assuming a constant rate of evolution, the time of their divergence can be calculated as **molecular divergence**. Example of a creature evolved according to the clock 580 mya, yet fossils of the creature were first discovered in 385 million year old stratum. The discrepancy may reflect a lack of fossils, or a misunderstanding about rates of evolution in a scientific prediction. These predictions may not be trusted. Concept time events are estimated times determined by molecular clock. The results of different tests of age may be millions of years. This method is used in the age of a group of organisms without relying on fossils and how much time elapsed since populations diverged.

Matching rock stratum can be dated. Recording of carbon ash (ca) test dating may have an error tolerance of plus 10% or more of the time value. Most animal fossils can not be dated.

Radioactive materials decay has a known rate. All living cells contain radioactive carbon 14 in proportion to the amount in the atmosphere. When a cell ceases to absorb radioactive carbon the quantity trapped within bone, wood, or other genetics begin to decrease at a known rate through radioactive decay that remains. The uncorrected errors tend to fall increasingly short of the actual age line. Basis; an assumption of a cyclic event of tree growth produces one ring annually regardless of its environment, for the tree growth rings of a 6270 year old Bristlecone Pine, effective dating of uncorrected radiocarbon dates is revised for a more actual date. After, 1000 years the radiocarbon dates divergence is plus 250 years, at 5000 years it increases to plus 1500 years. This makes **Mesolithic** stone tombs in Spain older than they actually are. Africa's Sademan volcano eruption 5.2 mya deposited seams of ash tuffs of silica glass. The glass is not radio-metrically datable due to the lack of minerals.

DNA degrades over time. Little Endogenous DNA remains in a bone 10 thousands of years old and is a tiny fraction of the total percent of DNA. DNA in a specimen fossil have from 5 to 1% contaminates from bacteria, minerals, elements, and others. When there are short strand fragments or a missing side of the double helix it greatly reduces the amount of material to work with. DNA does not preserve well in warmer climates or hot exposure to the environment. DNA does not date physical fossil remains. DNA for interbreeding 50 mya was CAATTTATCG. No interbreeding divergence of two groups with mutant gene changed 25 mya CAATT G ATCG and CAATTTATC T. More mutants built up in both groups so there is no interbreeding causing them to be distinct species CAAT CG ATCG and CAATTTAT TT.

Stereo-lithography computerized guided laser replicates the missing shapes of a skull by using plastic resin replication shaping of a art concept model. Computerized digitizing imaging creates a computer mirror image of a fossil by using the combination of CT-scan and 3-Dimentional Digital Technology for sculpturing.

Computerized Tomography CT uses thin X-ray beams creating a series of high resolution images from multiple angles transmitted to a computer. It creates a 3-Dimentional image enlargement and is rotational on a computer screen observation. Radiation exposure is minimal to high.

Micro Computed Tomography MCT takes thousand of CT images of fossil and digital fragments rebuilding a virtual fossil by making a mirror image representation of the opposite side of a face that was missing.

X-Ray Scan emits a broad beam that passes through the object being scan displayed on a film panel that casts shadow-like images. Fluoroscopy produces a continuous image movie. There is minimal and medium exposure to the object.

Magnetic Resonance Image (MRI) it used magnets and radio waves to create detailed images of tissue. It does not emit radiation.

Isotopic nitrogen composition analysis of teeth is used in the determination of diet.

ORIGIN LIFE SURVIVAL

A moment in time, when life began,
Organic micro-molecule cells developed, reproduced,
and adapted to the environment.
Relation Theory Natural Selection was the evolutionary process.
Larger micro-organisms, creatures, animals;
mammals, primates; apes and man evolved.
Food was the first challenge to resolve.
Bipedal walking and neurological DNA mutations in the brain developed.
Thinking, reasoning, and behavior changed.
Technology, tools, and social culture was developed.
Hunters and gathers now domesticate plants and animals.
With this nourishment, survival of the most fit will thrive.
J. Norman Wilkins

ORGANIC MICRO-MOLECULES

Earth had no ozone outer layer atmosphere protection 4.15 bya. Surface temperatures fell slowly. From the emergence of simple beginnings complexity evolves. Life evolved from organic micro-molecule process by natural selection. Life may have evolved introduced to the Earth from impacting rock matter from space 4 bya. Hydrocarbon residues in metamorphosed sedimentary rocks, in a single instance of such a condition, were the first organic micro-molecules. Single cell aquatic **Prokaryote** were organic micro-molecules (plant life) formulated as a chemical compound from elements of carbon, nitrogen, and oxygen as Amino Acid and Ribonucleic Acid (RNA) protoplasm necessary for all life. This single cell nucleoid was enclosed in a membrane and had a flagellum tail for mobility. Single cell form evolved tolerating the lack of oxygen 3.8 bya. It metabolized light energy from the sun. It may not have survived the early Earth violence as the first life form. This life could not reproduce. When this life was exposed to ultraviolet light it was destroyed.

Micro-Cellular Organism

Cellular Organisms are any living thing that is capable of sensation and motion in its complexity of its composition. **Thing** for our purpose, is matter with energy which exists as an individual distinguishing entity. The environment and climate changes had a great impact on the evolution of organic micro-organisms, "**Microbes**". Micro-molecules and RNA compounds are the origin for primitive photosynthesized micro-organism of bacteria and alga in the oceans. The fittest microbes thrived, while others decayed. There are two molecular genetic cellular challenges; grow strong fast and fit in times when food abounds, and decay in times when calories disappear. Surviving starvation is a key to life.

Many numbers of microbes photosynthesize in warm shallow water and evolved by forming mounds known as stromatolites, processed their own food, and thrived on light energy from the sun. Other microbes lived on the ocean floor near mineral chemical submarine erupting hot water mineralvents. Both microbs preferred an anoxic low oxygen condition. It forced them to adapt by retreating below the sediment and the others to the depths of the ocean vents. Oxygen levels slowly increased. Over time oxygen tolerant photosynthesizers inherited the Earth. Primitive photosynthesized micro-organism in the oceans released volumes of oxygen into the early atmosphere 2.7 bya. As the atmospheric ozone shield developed it had important evolutionary effects on the forms of life.

Both microbs produced oxygen (O2) as a waste biproduct that helped to form a primative ozone layer in Earth's upper atmosphere, blocking some of the Ultra-Violet (UV) light, and they changed the acidic oceans to a more alkaloid about 3.5-2.7 bya. These microbes had no specialized structure to reproduce.

Eukaryote organisms were multicellular complex cells containing a nucleus in a membrane 2.25 bya, having specialized structures, indicating the cell could reproduce. Fossilized embryos were discovered dating to 1000 million years ago (mya). Eukaryotes multicellular complex cells contained a nucleus with genetic material DNA within a nucleus membrane as red alga Bangiomorpha 1.26 bya. The many cells of these microscopic filaments show some specialization and structures, indicating they could reproduce into clustered growths becoming larger. There is an important message in the exchange of genetic material indicating it was not the only way to reproduce. Multicellular life could increase in size beyond microscopic with specialization of certain cells for certain tasks within the organism. The first silica skeletons in unicell alga formed 700 bya. Fossil embryos were discovered opening the way for larger and more diverse organisms 600 bya.

Over time the atmosphere and climate changed, the surface temperatures fell slowly and oxygen levels rose slowly. A different cellular life formed from micro-molecule elements of carbon, nitrogen, and oxygen (amino acid) combined to form the same geochemical **RNA** compound. The RNA applied geochemistry (energy) supporting the growth in the nucleus of the cell forming the heditary double helix **Deoxyribonucleic Acid (DNA)**; tightly coiled 48.74 inches or 2 meter long strands. These strands are the building blocks of life and the essential components for all living matter. Molecular chains out compete other molecular chains with the most fit usually winning the struggle.

The basic material within the DNA are **chromosomes;** microscopic rod shaped material of which the chromatin separates during the process of mitosis; the indirect method of cell division when the nuclear chromatin is formed into a long thread, which in turn breaks into chromosomes lengthwise transporting the genes conveying hereditary; characteristics inside the cell nucleus transmitting the hereditary patterns of the specie, its unique traits, and is the genetic code or **genome** needed to map each unique cell.

Reproduction in the female; each cell contains a small unit of protoplasm, a nucleus, in an enclosing membrane. **Metabolism** takes place inside the cell membrane, not the nucleus. The **Mitochondria** cell organelles, one for each cell, processes geochemical electrolites as released energy for cell growth. **Anabolism;** assimilates food processed into **protoplasm;** consisting primarily of water, proteins, lipids, carbohydrates, and inorganic salts, and **catabolism;** the used protoplasm broken down into simpler substance, or waste matter. The most fittest of cells survive.

These multi cellular cells start to multiply when the X (female) and Y (male) genomes mix inside the cell nucleus of the female mtDNA of X. Male and female with close genetics exchange their genes and supports the reproduction growth within the female and is passed on to the next generation young with minute changes between two compatible species by natural selection. Twenty three pairs (46) genomes of heredity, two compatable sets of 92 genomes are the building blocks for each unique life within the cell nucleus adapting to its environment. Specialized stem cells grow organs, bone, and others reproduce cells continuously replacing other cells. Oxygen-carrying red and white corpuscles heal injuries and fight infection as part of the immune system.

New mitochondrial DNA (mtDNA of X chromosome) is passed down only through females without the mix of Y male chromosome. It is the way of tracing ancestral information through females who had daughters. Humans all descended from one female to another female. Female mtDNA traced a common ancestor who originated in an area near Cape Town, South Africa, 117 thousand years ago (tya).

Genetic evolution mtDNA of X includes 36 genes inherited only from the mother and are the only genes traced that are linking ancestry to population history. The male passes on the 'Y' chromosome package and the female passes on the gene package containing the full gene complement from female to female. Females who beget only males dead end the ancestry linking, since the male cannot pass on a full gene package.

Comparing mtDNA and "Y" chromosome male equivalent is passed on to sons. The Y shows the change of polymorphisms over time combined to form haplotypes. **Haplotypes** of Y chromosomes form is the male equivalent of female mt DNA of X passed on from father to a son, known as polymorphism (miscreatants).

Any differences in mtDNA can be interpreted as mutations accumulated over time, which serves as a molecular clock, making how much time has lapsed since populations from different geographies have shown variations.

Comparing mtDNA chromosome from various populations, gives us a rough idea of where and when groups diverged on their ways in the great migrations.

Genetic difference between two species are calculated into two codified DNA's: Actual genes; over the genomes as a whole is a % (0.6) difference, due to a large sequence being deleted and others inserted changes having a much larger % (5) difference; non coded DNA difference between species is a % (1.2). This example is between Human and Chimpanzee relation. Three billion pairs of nucleotide molecules are considered to be 99.9%

similar among all humans. The other 0.1 % are the differences in traits making up each unique individual.

The differences in time between two groups separated from a common ancestor, the greater the difference in their DNA. The difference can be calculated.

One **Gene** contains two hundred fifty thousand human cells. Mutants of the DNA change the physical traits for growth, health, and development. Each genome will have minute changes in genetic reshuffling "Markers" that takes place each time the female and male DNA combines in the female nucleus, creating new life, and creates a new evolved related descendent preserving the telltale variations of genus, species, and variants passed on to the next generation. Occasionally, time will at random produce mutant or markers in the genome genetic function of stretches from adaptation and environmental changes between species and variants transformed as a new and different related descendent.

Occasionally, there are miscreants of a genome link when a trait changes places with other traits. Most are normal, but, some may be abnormal for growth, health and development. Identifying these traits from blood and bone can provide information about our ancestry and physical health markers genetically passed on from one to another.

Miscreants change **Traits;** any distinguishing quality or characteristic of genetics inherited from ancestors; genetic traits encoded into the genes, the hereditary units of DNA in the chromosome of our cells are the genetic profile changes, geographic isolation, and mutation over time producing new genera.

Aberrant; a cell errs during division instead of the 23 chromosome pairs having accurately duplicated, bits of one number and another number change places creating an aberrant, diverting from what is normal, bearing a mutant miscreatant gene.

DNA Karyotype; human genome containing 46 chromosomes (23 pairs), two chromosomes inherited from ancestors we shared with other Primates are fused togather to the number 2 chromosome.

Over time cells adapt to environmental changes in competitive survival with an occasional imperfect mutant; those inheritable characteristics that differ from those of the parent miscreant, causes a person not to believe or to be surprised. New genus and species evolved, adapted, and changed the interaction of the molecules and minerals of the cell structure to their environment. Finding these **marker** events of a gene resulting from mutant changes in form variation in some inheritable characteristic of two individuals DNA indicates they share the same ancestor tracing their connections.

DNA is chemically analyzed. Testing and data reporting is preformed by technicians using digital computers. The data from analysis is compared from many tests in determining relationship of individuals. The genetic coding strings of alpha designated (ACGT) sequencing consist of 3 billion pairs of nucleotides of molecules making up 99.9 % for all humans. The other 0.1% is the differences in trait markers making each unique individual.

Molecular biology DNA sequencing traces racial descendents from skeletal and linguistic features. They identify and document the changes. When and where the fossil lived, intermixing of the descents ancestry, migration, environment, and adaptation tells a story of their origin. The ape, monkey, chimpanzee, gorilla, and human blood protein were genetically tested and compared for differences in genes. Every drop of human blood contains a history book written in the language of our genes. Genetic alpha designated sequencing of the genome DNA code; A for adenine, C for cytosine, G for guanine, and T for thymine makeup of one human Homo genus is 99.9 % identical throughout the world population. Alpha codes with one or more suffix numbers have a specialized genetic marker detail of the basic code. The codes remaining, 0.1% are the snippets of DNA differences responsible for unique individuals. There are some codes that do not serve, at this time, any known function.

DNA string of genetic genome code of H. sapiens (CGTA) and H. neanderthalensis sapiens MCIR "Neanderthal" are separate species of Homo sapiens. However, some did beget. Some H. sapiens of today may contain a small amount of Neanderthal genome code, as much as 2.5% of the 99.9%. Some H. sapiens escaped European Neanderthal and African influence by living outside of those population areas.

From the Homo sapiens twenty three genomes, mutation variant DRD4 helps control dopamine brain messengering important in learning and reward. DRD4-7R makes these humans more likely to take risks; explore new places and ideas, relationships and sex opportunities, change, and adventure. Ancestors with the 7R restlessness behavior action thrived in the environment and are tied to those who migrated.

The way humans live their life determines the ageing of their cells. The body requires full range motion. Exercise stimulates the need to eat. **Motion** by exercising and **emotion** by objectively thinking are the signaling system that tells cells to grow. Without it the cells decay during their endless cycle of replacement. Over time, as we age, our oxygen necessary for life degenerates and causes the cells to oxidize. Antioxidant supplements slow the process of cell oxidation. Only key stem cells in organs and in the brain survive long term. Our daily cell replacement accounts for 1 % of our body and every three months we have a whole new body.

The first sea dwelling **soft bodied animals** evolved 580-543 mya. Their evolution may have been from the melting of the last snowball event 800 mya. Aquatic life exploded in the oceans. Major groups of marine **invertebrates** lacked a backbone. Fish like animals with the beginnings of a backbone called a **Notochord** had evolved by 525 mya. Fifty percent of this genera of marine organisms became extinct 443 mya due to the ice age 445 mya. Wingless insects formed 405 mya. There was diversification of plant life, fauna and the first forests covered the low wetlands. Fish evolved four limbs, making them the first tetra pods who ventured onto land to feed and back to the water to breed 370 mya. The first **Cynodonts warm blooded vertebrates** evolved 255 mya. Amphibians and reptile-like forms evolved. They laid a membrane-covered amniote egg on dry land. The dominate reptile groups; dinosaurs on the land, ichthyosaurs in the sea, and pterosaurs in the air 230 mya. This group diverged into dinosaurs and others into **mammals.** The primitive mammalian **Purgatorin** evolved as a warm blood vertebrate with placental births suggesting the genetic exchange by male and female forms 124 mya.

Ancient Creatures

Many **Ancient Creatures** evolved and thrived in the ocean and grew in size 550 mya. Divergence of the genera of vertebrates was too many to be classified 540 mya. The first endothermic emerged from the ocean to land as adapted warm blood vertebrates and were more active as **Cynodonts** 255 mya. The endonthermic with cold blood relied on the environment to determine activity as the temperature increases.

Creatures existing of any kind <u>are not mammal or primate</u>. **Kind** refers to the natural group or division of creatures. Creatures changed drastically demonstrating evolution is the building on what had been built on before by Natural Selection. Species had to adapt to survive, breed, and pass on those special characteristics to the new generation.

THE OLD WORLD

Pongidae

Pongidae, "Pongids" (PON) evolved differently in divergence, form, transformation, and transition. **Eomaia (EOM)** of many forms in physical changes as variants were too specialized to be classified. The first Encyclopedic of Life by Carl Linnaeans system for genera classification; 1st word letter is capalized of the genus and the 2nd word letter for specie is not capalized. Linnaean a Swedish botanist cataloged the name Anthropmopha meaning; a human form or man shape. He changed the order to Primate to appear first in the class of Mammalia (mammal).

The first birds Archaeopteryx evolved 150 mya. The first flowering plants, feathered dinosaurs and placental mammals evolved 124 mya.

The earliest **Common Primate (CP)**, were warm blooded vertebrates having placental births. **Common** for our purpose is shared by belonging equally to all primates. **Primate Placental Mammal (PPM)** animals of the highest order of CP diverged as **Common Ape (CA)** developing their unique traits over millions of years. **PPM** animals evolved into many divergences of primates 124 mya. **Mammals** are any of a group of vertebrate females who have milk secreting glands for feeding their young. Dominate placental mammals gave birth to bigger more advanced young during the early 85-70-65-54.8 mya. These arboral animals had larger brains, lived in social groups as adult males, females, and their young.

Primate Placental Mammals included the gorilla-champanzee and monkey-ape forms and were classified differently from other mammals starting to evolve 85 mya. The new order of mammalian primate change started about 40 mya. These groups populated Africa, Asia, and South America. There was a reorganization of New Order Primates geneus divergence into four different and separate genus 30-20 mya. The 1st group of gorilla diverged from chimpanzee 27 mya. The 2nd group of monkey diverged from ape as **Early Common Ape (CA)** 20 mya. DNA evidence supports Gorilla and Chimpanzee are related, but are not ancestors to CA. However, the four genera are all related to CP.

The Early CA survived in South America, Europe, Asia and Africa 19-17 mya. Early CA walked on their palms and had not been through the evolutionary gorilla-chimpanzee ape phase of knuckle walking. **Later CA** was begat to or closely ancestral to African CA lived in southwest Kenya and southeast Ugonda Lake Victoria area 16 mya.

Purgatorius (PUR) were the first primate mammalian living in North America 65 mya. It was a monkey-like primitive primate. PUR evolved as a shrew-like insect eater. PUR were followed by the primates Tarsier and a Lemur-like emerging in South America, Europe, Asia, and Africa 55 mya. Some of them survived the K-T Asteroid Event and the Deccan Lava Eruption Event 65 mya. These events caused a global ice age.

The first **Plesiadapi-forms** and **Altiatlasius** were ancestors to others like the Tarsier living in Morocco, Africa 56 mya. They were squirrel-like tree climbers. They were too specialized to have been begat to modern primates.

Plesiadapi; Strepsirrhines and Adapid lived in Europe, North America, Africa, and Asia begat to the Lemur. **Haplorrhines; Omomyid** (OMD) lived in North Africa and in all of the above areas 54 mya. OMD were early evolving and were rapidly diversifying. OMD were the ancestors of Tarsiers and monkey-apes 54 mya.

Prosimians were primitive tropical primates like tarsier and lemur, begat from Plesidapi-forms 50 mya.

Anthropoids evolved from OMD and lived in North Africa and differ from Prosimians (Lemur) 40 mya. They also, lived in Egypt, China, Myanmar, and Thailand. The **Anthropoid eosimias (ANEO)** were the best known and lived in Myanmar 40 mya. **Aegyptopithecus (AEGY)** lived in Fayum, Egypt 33 mya.

ANEO features: skull; small sagittal crest with a rear horizontal extending nachal crest, short muzzle face with nostrils, protruding forward front teeth with long curved canines (ever growing incisors) and eight flat molars, and large eye orbits with bony back walls. They were arboreal in the forest canopy.

Ankaraopithecus (AK) lesser apes like Gibbons and Siamangs evolved in Turkey and at the Hungarian site of Rudabanya, located on the western flank of the northern Carpathian Mountains in Hungary. Not an ancestors to man.

The climate was warmer and wet 23 mya.

Primate populations reduced in Northern Europe and Asia.

Proconsul africanus (PA) and **P. heseloni (PH)** variants from the 2nd old world group of monkey-apes lived in southwestern Kenya on Rusenga Island, Lake Victoria and Heseloni, Africa 23-19.4-18 mya. PA and PH are the same specie. The New World common Apes diverged as Great Apes and diverged as specialized apes.

PA and PH features: the skull; had a monkey looking face and head, 167 ml brain, and a narrow short snout and muzzle. Its forelimbs became more like arms and aftlimbs more like legs. These apes were arboreal primates, climbed trees, and walked as a quadruped. It was tailless. PA is related, but are not ancestors to man. Some of these members grew in size. Another PA variant lived in western Koru, Kenya 20 mya.

THE NEW WORLD

The new world of time started about 30-20 mya. It has many gaps in the fossil record from the Common Ape divergence to the specialized Super Family of the Hominidae Australopithecus ape and **Homo** genus of primate that includes modern man. Monkey populations were decreasing. Some of the ape populations were developing. Fossils disclose clues about distinction and variation of bone used to group populations distinct enough to propose new genera and species. As the fossil record grows species were redefined, grouped, or eliminated. There were signifient developmental differences in the skeletal remains between ape, apeman, manape, and man.

Apes became more diversified and were diverging, transforming and transitioning to other forms in South America, Northern Europe, Asia, and Africa by 22 mya. There was a reduction of apes and monkeys in Europe, Asia and Africa. The apes survived 19-17 mya. **Apes** were diverging into new genera; **genus** is the main subdivision of the primate family and includes one or more **species**; having the appearance, shape, and distinct kind of animal distinguishing characteristics of the genus like Australopithecus afarensis or Homo erectus variants; the degree of change or difference in some way from others of the same kind.

Primate Apes

Learned survival behaviors were used by small brain apes before any major changes to others who gathered and hunted. They used their sensory processing to mentally map a location where they found food. When they want to return to the food source location at a later time, they mentally recall and return to the food source. The brain provides the trigger for motor action to the muscles to perform body tasks. They adapted by using stone tools to crush hard food. Time passed and they learned primitive capture of prey and scavageing of other animal kills supplementing meat into their diet. Later they learned co-operation by the members in the hunt, to drive prey from the rear toward other member's, who ambushed the prey in front by blocking the escape, catching and securing it before it discovered an alternate escape. They hunt to eat and kill because it makes feeding more manageable. They made tools to support the hunt and learned to adapt from experiences in surviving.

Some of the **Primate Ape** variant forms diverged and were classified. **Dryopithecus** (DRY) evolved in Europe and Eurasia 17-9 mya. Four other species lived 12 mya. DRY is begat to Gorilla-chimpanzee and Ardipithecus.

DRY fontani (DRYF) variant meaning, "Oak Ape", lived in Europe on the north flank of the French Pyrenees at Saint Gaudens, France 17-14-9 mya. DRYF had a combination of primative and advanced features; chimpanzee size; high built skull; well developed brow ridge, and small thin enameled teeth. It had long fingered hands and arms adapted for hanging in trees and legs adapted for climbing trees and walking along the top of branches eating tender leaves and fruit.

DRY **laietanus** (DRYL) and **crusafonti** (DRYC) variants lived in Valles Penedes, Spain. DRY **brancoi** (DRYB) variants lived in Mariatal and Ebingen, France, in Baden and Wurttemberg, Germany, and in Euroasia Rudabanya, Hungary on the western flank of the Carpathian Mountains, and Udabno, Central Russia between the Black Sea and Caspian Sea 12-10 mya.

The climate was cooling in Africa, deserts were formed, and forests disappeared changing to grassland 16 mya. Claims were made **DRY Africana** (DRYA) variant was discovered in East Africa and throughout the countries of Ethiopia, Kenya, and Tanzania living in the Great Rift Valley System 10 mya. Related, but, are not an ancestor to man.

The **Sivapithecus** (SI) diverged from the Asian Orangutan evolving in China and Southeast Asia 14-10 mya. They were similar to their Asian ancestor Orangutan. SI lived in the Sivalik Hills of the Hemalayas. It was arboreal using their arms to move under the branches, walking on the branches palms down, and had long muscular hands and fingers. SI had a thin bone skull; well developed brow ridges and small thin enameled teeth adapted for soft food diet. The lower jaw with teeth was discovered at Haute Garonne, France.

Sahelanthropus tehadensis (SAT) evolved about 7-6.9-6.3-6 mya. It was named, "Toumai", meaning, Hope of Life, in the Gorman language. There was a discovery of a skull in the southern Sahara region of Bahagal el Ghazal Toros-Menalla, in the Djurab Desert ancient shoreline of Djurab Lake, Chad, Africa. It was 1500 miles west of the Great Rift Valley of east Africa.

SAT features: skull; appeared to set on top of a spine, the brain was 370 ml, very prominent brow ridge, flat and vertical face, with heavy enameled small teeth and canines. SAT had traits and some adaptive human features, but was not hominid.

The 1st **Indeterminate** jaw fragment of a primate ape may be a variant between Ar. and Australopithecus (A.) discovered at Lothagam, South Lake Turkana, Kenya, Africa, in 5.6 million year old stratum. It was not hominid.

Babrelghazali (BAH) "Abel" a bit of a jaw was discovered from the Djurab Desert, Djurab Lake ancient shoreline, at Toros-Menalla, Chad, Africa in 3.5-3 million year old stratum.

Primate Great Apes

The New Order PA diverged as **Great Ape** genus, species and variances. The European, Asian, and African Great Apes were developing during this time. Some of the **Great Ape** variants grew in size. The Gorilla, Chimpanzee, Monkey, Common Apes, and others diverged and evolved as many different species and variants. Some apes had strong long curved fingers and palms, stiff joints in the hands supported their weight on the knuckles when they walked on the ground "knuckle walking". The knuckle walking was an adaptation for the gorilla and chimpanzee and was an evolutionary requirement. Other apes walked on their palms and had not been through the evolutionary gorilla-chimpanzee ape phase of knuckle walking.

Archaic Chimpanzee (ACHI) diverged from gorilla-chimpanzee 27 mya. ACHI was a primitive tool user with complex social behavior. Males dominate the females brandishing their canines to intimidate other males. ACHI wrists were stiff allowing it to support weight on its knuckles on the ground. Related but is not an ancestor to man.

Kamoyapithecus Ape (KA) lived in Africa 27-24 mya.

Ramapithecus wicker (RAMW) was renamed from **Kenyapithecus wickeri** (KENW) variant evolved as a Chimpanzee Great Ape and lived in the Rift Awash Valley System, Ethiopia, Africa, and at the Fort Ternen site in Kenya, Africa.

RAMW were similar to the chimpanzee and size with features from the fragments of a skull and jaws with teeth. It had long arms, short legs, and a thick waistline. It lived on the tundra grassland and woodland, not in a forest and was discovered in 14-12.5 mya carbon ash (ca) stratum sediment.

Late Common ape diverged as a Great Ape variant transforming as a specialized genus **Archaic Australopithecus** 8-6 mya. Many Australopithecus (A.) species and variants lived at the Lake Tukana border between Kenya and Omo, Ethiopia 8-4.9 mya. As time passed, they transformed and transitioned to other forms of their kind.

Chimpanzee (CHI) transitioned from ACHI 7-6.5-5 mya. CHI evolved as a Great Ape with some adaptive traits. They shared common relationship to other Primate Great apes. CHI and humans are genetically different, but are closely related. CHI shares 96% of DNA with other hominid apes, but still walked on their knuckles. Late CHI transitioned from CHI 3.5 mya. Modern CHI transitioned from Late CHI. Their DNA sequence difference was 98.7 % and 1.2 % separation from humans 800 tya. Modern CHI variants evolved and exist today. There is no known evidence of interbreeding with humans, primarily because of their relation difference. An explanation of Genetic differences between two genera related forms can be reviewed in the DNA part of the text.

Orrorin tugenensis (OT) was discovered at the Lukeino Formations in Cheboit, Kenya in 6.2-5.9 mya stratum. It evolved and adapted from a primitive ape. OT was begat from Common Primate (CP) variant 85 mya. According to Tugen legend OT is the original spirit that settled in the Tugen Hills of Kenya, from the Afar language, known as the, "Millennium Man" or man of the Tugen Hills. OT features: skull jaw bone with teeth, arm, finger bones, straight shaft thighbone, with a thick boned upper ball joint fitted into the hip socket. It was far from being hominid.

For our purpose our ancestory evolved from the Great Ape specialized Australopithecus; transitioned to apeman, manape, and Man. A. quadrupeds roamed east Africa of the Great Rift Valley System, including the Olduvai Gorge flanks. The North African woodland areas were decreasing with open areas of savannah. The climate became increasingly arid as the Earth slipped into an ice age 4.5-4.3 mya. These plant eaters and tree dwellers survival was greatly dimished and many became extinct.

Australopithecus anamensis (ANA) variant was a specialized ape (of the lake) evolved in equatorial eastern Africa in the west rift margin at Lake Turkana, Kanapoi and Allia Bay, in northern Kenya, Africa and at Aramis one mile northeast of Hadar, Ethiopia, living in these areas 4.9-4.2 mya.

ANA features: small ape skull; jaw jutted forward, heavy brow, flared cheeks with strong muscles, and an erect posture. It walked bipedal allowing for less body surface to be exposed to the hot sun in those areas where it was very warm. The less surface exposed to the sun, helped in body air cooling. Great Apes had larger brains, lived in complex social groups, more flexible, manipulative, and agile. Most of them had a large stockey body, short arms and legs, and evolved without tails. The bipedal and palm walking was an adaptation. ANA was a palm walking quadruped ape and had not been through the gorilla-chimpanzee ape phase of knuckle walking. It walked as a primitive bipedal 4.2-4.1-3.9 mya.

Ardipithecus (Ar.) genus fossils were discovered in the Middle Awash Valley, Central Awash Complex, in an area around Aramis and Amba East, Africa. Aramis was one mile northeast of where Ar. fossils were discovered at Hadar and Ethiopia's west margin of the Awash Basin in stratum dated to 4.4 mya. Ar. diverged as a specialized ape form of Chimpanzee (CHI).

The two species of Ar. represent two arbitrary points in a single evolving lineage with no clear dividing line between them. Ar. ramidus kadabba (Arrk) and Ar. ramidus (Arr) differences did not change over the passage of time.

Distribution of weight over the feet is an efficient way of walking along tree branches for an ape. The apes already had the rudiments of bipedal when they ventured out of the forest. These Great Apes had adapted some transitional traits for primitive bipedal walking on the ground behavior in combination with primary quadruped "palm walking" life in the woodland trees 5.4 mya. There is no proof they were bipedal. It diverged as specialized relict specie variant of the chimpanzee carrying its primitive and advanced traits with it into extinction within 200 thousand years.

Ardipithecus ramidus kadabba (Arrk) evolved and lived 5.8-5.2 mya in the badlands of the Central Awash Complex, at Aramis, Ethiopia, home of the Ar. Arrk variant appears to be a chronological species of Arr. It had a small brain and reduced canines. The discovery of a molar tooth of a young adult was located in a dry lake.

Ardipithecus (ground floor) **ramidus** (root) (Arr) "Ardi" evolved and was named from the language of Afar about 4.5-4.4-4.37 mya. Arr fossils were discovered within the west margin of the Awash Basin near Hadar, Ethiopia in 4.4 mya stratum. Fossils of Arr resembled Arrk. It lived in the western rift margin of the Nubian Plate of the Afar Basin, Middle Awash Valley, southeast of Aramis, Ethiopia in an equatorial warm climate woodland forest environment. Arr behavior was better at cooperating in large social groups, less likely to become a big cat or hyena next meal. It was feast to famine to survive. The savanna environment change created incentive to find food. They obtain food that was harder to find. The change of diet drove the development for larger brains.

One hundred twenty five fossilized fragmented pieces of a female with a muscular body were discovered. She was 4 feet tall, weighing 110 pounds, and was the size of a chimpanzee. She was close to becoming hominid and is related, but is not an ancestor to Australopithecus (A.). Arr developed in separate ways and in different directions from A.

She lived in the trees of the closed canopy woodland. She gathered and ate food in the forest from the trees; hackberry, fig, and palm trees. She slept in a nest in the trees. On the ground she ate plants, eggs, grasses, tubers, roots, seeds that grew on the savanna, and small creatures and animals. Arr lived overlapping monkeys, kudu and peafowl that prefer the woodland to open grassland.

Her primitive skeletal arrangement was shared by other hominids. It had a stiff ridged back for climbing. The skull; small brain, ape like muzzle, sloped face, closely spaced eyes, a bony brow ridge, and had small canines. The opening for the spinal cord in the base of the skull was positioned far forward indicating the skull was balanced on top of a near vertical backbone. Her lower lumbar vertebral of the spine are not attached by ligaments to her hip blades, as they are in modern apes.

She could balance upright on one leg at a time, while walking on branches, and on the ground with a shuffle. The plavis had large iliac bones. The upper pelvis was shorter and broader than apes. Protrusions on the inside edges of the palvis were added during the adaptation reinforcement. This enlargement allows for the attachment areas of the gluteus muscles that stabilized the support of her hip joints, allowing walking without learching from side-to-side. The lower pelvis had large powerful hip and thigh muscles that would have made it difficult while on the ground to run fast or far without injuring her hamstrings.

She had long arms. The wrist bones are a more flexible structure, while using their palms moving along the upper surface of the tree branches. The hands had long fingers and short palms fitting together in a way that allowed the hand to bend far back at the wrist. The wrist bones had a more flexible structure allowing her to walk on her palms in the top of trees branches like a monkey, not an ape. This indicates she did not go through evolved wrist knuckle walking after coming down from the trees to live on the ground.

She had long legs. The bones of the toes looked like a primitive variant of Arrk. In apes the foot joint bone medial cuneiform surface points in a different direction. These are not the feet of a biped. The extra bone in the wide diverged opposed splayed big toe pulls away from the grip against the other toes allowed it to be kept more ridged while walking to grip branches or on the ground, giving the foot a dual role. In apes and monkeys the os-peroneun is a small bone that helps propel the opposed big toe in upright walking, retained in hominids, kept the bottom of the foot ridged without an arch. This rigidity enabled Arr to walk upright on the ground using its four aligned toes to provide the leverage, "toe off", that propels bipedal stride. This combination of traits was different allowing Arr to live in a two world environment. She could walk upright, however, could not walk over long distances. Her big toes would have made it difficult getting around on the ground.

Chart 1.
Phylogeny Old World Pongidae Common Primate Ape
New World Homindidae Great Ape 124-4.5 Mya

Primate Population Loc.*	Species variant
ET	4.5 Ardipithecus ramiud (Arr)
ET KN	4.9 Anamensis (ANA)
KN	5.6-5.5 #1 Indeterminate
ET	5.8 Ardipithecus ramidus kadabba (Arrk)
KN	6 #2 Indeterminate
KN	6.2 Orrorin tugenensis (OT)
	7 Chimpanzee (CHI)
CH	7/6 Sahalanthropus tehadensis (SAT)
ET	8 Australopithecus (A)
SP	9 DRY laietanus (DRYL) and DRY crusafonti (DYRC)
ET TA	10 Dryopithecus Africana (DRYA)
FR GR HU	11 Dryopithecus brancoi (DRYB)
	14 Ramapithecus (Kenyapithecus) wicker (RAMW)
FR	14 Sivapithecus (SI)
FR	17 Dryopithecus fontani (DRYF)
NEW WORLD HOMINIDAE	
	20 Early Common Ape (CA) PON
KN	23 Proconsul africanus (PA) P. heseloni (PH) PON
	27 Kamoyapithecus (KA) PON
	27 Archaic Chimpanzee (ACHI) PON
NEW ORDER COMMON PRIMATE	30-20 Gorilla-Chimpanzee diverge as Gorilla and Chimpanzee
NEW ORDER COMMON PRIMATE	30-20 Monkey-Ape diverge as Monkey and Early Common Ape
HU	30 Ankaropithecus (ANK) PON
	33 Aegypithecus (AEGY) PON
MY TH	41 Anthropoids eosimias (ANEO) PON
CH EG	
	55 Plesiadapi-forms (PLA) Haplorrhine: Omomyid PON
	65 Purgatorius (PUR) PON
UG KN	85 Common Primates (CP) PON
	99 Eomaia (EOM) multi-forms PON
	124 Pongid (PON) Placental mammal warm blood vertebrates
OLD WORLD PONGIDAE	

TIME MYA   124   99   65   55   40   30   20   18   17   14   7   6   5   4.5

* See Primate Population location by species chart 4; African, European, Asian

Origin of Primate Family

There are questions needing answers concerning the actual age of fossils. The direct actual analysis of accurate and factual data may be unclear or reported incorrectly. There are risks for errors in what is not proven in the examination of fossils and bones by respected anthropologists in mutually contradictory interpretation, communication, and expert opinion leaving the task open with unattainable results for future investigation. DNA testing and analysis by scientist and laboratory specialists must take extreme precautions while handling specimens to ensure the probability of contamination and subjectivity is minimal as documented in their findings and reports.

ETHICS

The scientific analysis derived from fossils, bone, and stratum facts produce objective and subjective data. Science also uses positive and negative tolerances because it is not an exact process of determination. The processing accuracy is calibrated within ranges of tolerances and values used in a process, included are error skewed data resulting in negative and subjective analysis. The testing equipment must meet critical criteria specified in established positive and negative limited standards for objective analysis. How effective the standard basis is, determines how objective the critical results are and how much the error acceptable limits are from the zero true reference plus and/or minus tolerances. Some testing equipment may be experimental, research and development phases of refining the standard limits closer to the true zero bases. The data averaging using high and low limit values or assumption based on the value of related facts may not be reliable data. The importance of data values to the zero true reference may not be critical resulting in a margin for error and subjectivity.

Technology and the information digital data world continuously changes with more accurate ways to analyze data. Our curiosity about our past will probably outstrip our ability to collect its remains. We need to know how ecology and the social conditions of things occurred. We learn about the trials of life and society. There is a need to be honest and skilled with investigation tools used concerning known facts. There are risks for errors in what is not proven. Errors in the testing data base are critical to the highest potential of excellence above the standards in disclosed reporting. While we keep searching for the answers to distinctively human life there must be accountability for responsible expert testimony. Exactly how old that animal was is not as important as the origin and adaptable evolution. When there is subjectivity in the analysis or no proof, the creditability by those who wax the truth or impose their own spin on their findings, for their benefit, is questioned. Scientists name their discoveries and determine approximate age from surrounding stratum. Testing is not finite.

Facts and studies performed by the scientific community are still discovering new fossils. There will be events that are key details about the unknowns. There are risks in objective assumptions becoming even greater when there is subjective analysis. There are conflicting interpretations, assumptions, theories, and arguments about the anatomy of hominids and humans. Predictions from authorities are not to be trusted. They need to get out in the field and find what was predicted and then determine the facts about the prediction. There is a need to deal with the facts, guard what is said that are opinion, tell the truth, and prove it. There is no recourse for the truth.

Discovery is not finite. There are discovery disclosure errors in exhibits found in tectonic activities of erosion, faults, river wash, canyons, sediment, volcanic eruption, and fossils displaced to another location that may be different or unknown.

The chain of evolved ancestry is linked by gene mutation passed on to successive generations through DNA. It scientifically unlocked a better image of technology. DNA helps to explain life's Relation Theory Natural Selection.

ORIGIN OF THE PRIMATE FAMILY

The Origin of the Primate Family evolved from vertebrate primate mammal animals, specialized great apes, apeman, manape, and finally all man. These species and variants adapted to their environment and had mixed differences; diversity, divergence, transformating, and transitioning passed on to the next generations. They developed physical and mental changes. The mutation of the brain development of Modern Man over time resulted in strategic thinking and reasoning to plan and innovate. Anatomic Modern Man's brain developed with the frontal lobe with Mind-Brain behavioral thinking; cognitive traits, creative thinking, creative art, fully articulated speech and language to build on social culture, symbolic thinking, the ability to abstract and analyze the past, anticipate the future, and develop new tool technology.

Our family genealogy presents conflicting interpretations spanning possibly 255 million years of organic micromolecule, micro-cellular organisms, ancient creatures, mammals, primates, apes and man. Over time in the old world primate mammalians evolved as many divergences, transformations, and variants of the primate life. Dominate placental mammals gave birth to bigger more advanced young. **Common Primates** (CP) were the most highly developed order of mammal warm blooded vertebrates having placental births.

From the emergence of simple beginnings, complexity evolved. Homo sapiens ancestry is continually being recorded filling in the gaps from the early Common Ape (CA) divergence to the Super Family of the Hominidae specialized Australopithecus (A.) ape, apeman, manape, and man.

The **Culture** of groups of individuals lived together, learning from experiances in their environment, established their language, technology, customs, and beliefs transmitting that knowledge, skills, and abilities from that generation to another. There is proof and facts about primate fossils discovered in the sedimentary stratum of the Earth. Human origin was not created in a series of events by a spirit or by any conceived supernatural immortal having special powers over a person with everlasting fame. Inspired scular writings and teachings were the words, guidance, influenced, and the conversion and persuasion of those free willed believers and the faithful. There is good value in the tales and stories of common sense interpretations. **Spirit Theory,** belief and faith without the supporting proof of truth and facts about life's origin has no recourse. Scientific analysis, DNA sequencing, and technology provides a better understanding of Relation Theory by Natural Selection closer to the truth than Spirit Theory and the after life theory.

Living organic micro-molecules, over time, diverged to primate apes genus, species, and variants and are responsible for man Homo sapiens adapting to the environment.

Environment and climate changed the effects on evolution. Hominid ancestory has gaps in the overlapping and research interoperations that continue to probe the unknown. Over time, Hominids diverged from other quadrupeds to upright two-legged bipedal primates. There were other species and variants evolving contributing significantly to the biological adaptions to their environment, including Homo, passed on to their descendants of their kind. The East Africa Great Rift System tectonics divergence was 32-25 mya. In northern Egypt the Nubian Plate diverged northeast opening and created a dry depression with east and west margin that later became the Red Sea. The rift's south western margin diverged southwest at Eritrea, Ethiopia. The eastern margin opened at the mouth of the Golf of Aden, Djibouti, Somalia, forming the Afar Basin rift in Ethiopia, extended through Kenya, Tanzania, and included the Olduvai Gorge rift. The Somalian Plate diverged southeast opening and created the eastern margin. The formations cracked into slabs and were strewed about haphazardly.

The eruption of Sadiman Volcano gushed basalt lava and ash over the enormous Serengeti flood plains 5.2 mya. The eruption deposited thin seams of glass tuffs interlaced among the basalt and ash sediment. Over time sediment built up covering the basalt base, tuffs, as erosion slowly changed the landscape and environment. The volcanic silica glass was located in ridges of seams called Lusaka Tuft, "lion hair", in the Afar language. The glass is not radiometrically datable due to the lack of minerals.

Near the village of Herto, Ethiopia, the volcanic flow formed a natural dam, "The Bouri Peninsula", spans the rift, blocking the west and east margins. It forms the north boundary of Yardi Lake and the mouth of the Awash River flowing northwest to the Red Sea through the Awash Valley System Afar Basin.

Other A. and H. genera, species and variants evolved 2 mya. There were other species and variants who contributed significantly to the biological adaptations to their environment, including Homo. They evolved passing on adapted genes to their descendants of their kind.

Some did not adapt to their environment and became extinct. However, surviving A. with small brains transitioned with gene molecular mutations in successive generations, evolving as apeman with greater brain and more advanced human development challenges. There was transformation and transition of others as H. developed a bigger brain and adaptive traits from their environment. There were H. habilis, ergaster, and erectus in Africa before Imya. Others lived at Saldauha Bay Lagoon, South Africa, along the shoreline of the Atlantic Ocean, 75 miles west of Cape Town.

Gene mutants and a diet of meat eaters, developed larger brains, with greater thinking, reasoning, behavior, social culture, and new stoneage tool technology. Many were transforming and transitioning in Africa and evolved as apeman, manape, and man.

With a brain size 3 times in volume of an ape, early genus Homo became a better hunter and gather. From infancy essence through adolescence male children devoted their time to learning skills and social behavior. With these skills their initiative expanded the tool technology, culture, social behavior, creativity in art, and a language spoken for communication.

The senses, emotion, and stress played an essential part in how humans react in their environment. It was not until Anatomic Modern sapiens emergence with major neurological mutant gene adaptation of the brain changed cognition and development of new skills. They were vastly more complex. With the frontal lobe development of the brain neurological mutation led to a spoken language and cultural behavior. The Mind process; creativity, systematic thinking to confer an enhanced ability for reasoning; to plan, innovate, make decisions, and take corrective action. This reasoning is not shared with any other animal. All these mind-brain functions are needed in a civilization.

Over time the brain and mind developed to modern thinking. The Mind is our **consciousness** and does not reside in any one part of the brain. The mind <u>neurons</u> searches and communicates (bridging) over nerve sensors of the brain <u>receptors</u> through neuro-electro-chemical <u>synapses</u>. The Mind thinking process receives information from various senses, encodes, and maps the received information for long term stores in the brain in various areas for recall by the mind at a later time. The Mind may decide to process received information that is short term, is not signifient for brain action, and was not encoded for long term. Because we do tasks one at a time, multi tasking distracts, if there is time using a check list or writing down the add tasks helps any of five senses (sight, hearing, touch, smell, and taste) to recall and brain action for physical performance at a later time. Stay focused performing tasks. The Mind determines by answering why in making decisions, takes action on the decision and behavior, resulting from emotion and stress. The Mind reasoning with the brain retrieves prior events from the brain storage areas, sends trigger messages to be applied to the brain receptors nerves to perform motor action for physical body functions. Some body and organs automatically change continuously or function without mind-brain synapses.

Violence is not in our evolutionary ancestry or in the genes. The learned aggression and defensive action behavior attitude is used as a survival trigger for ill deed. They killed each other when they started to compete for power over others. Some resorted to cannibalization to survive. Some failed to survive adaptation to their environment.

Reasons for genus, species, or variant to cease to exist; there were no living or to few reproductive females of the species, the surviving males or female beget more often with other domonite species transformed or transitioned variant, no living variant species, differences in not adapting to the environmental resources, segregation into isolation by dominate species, competition for resources gives rise to convergence that became collision, provoked war, disease, and extinction.

THE HOMINID FAMILY
OF HOMINIDAE

Ancestory is continually being recorded filling in the gaps from the Common Ape divergence.

Hominids of or like man: manlike; are any of a **Supper Family of Hominidae** who were all two legged Great apes. Man's ancestry; as a specialized Australopithecus (A.) ape, known as, "Zinj", Zinjanthropus, transformed and transitioned as apeman, more than half ape with some advanced man traits. As generations passed, apeman transitioned as manape with more than half man advanced traits. Later manape transitioned as modern forms of man extinct and living.

THE FAMILY OF EARLY ANCIENT
HOMINID 4.4 TO 1.9 MYA

Hominid fossil evidence was discovered in stratum and sediment dated to 4.4 mya. They were from Africa's Middle Awash Valley, Afar Basin, and the African Rift System in Ethiopia and Tanzania. There were discoveries in South Africa dating to 3 mya. Alisera tribesman who were nomads living near the village of Aramis, Ethiopia, resisted the scientific teams advances at the dig site, because the team was disturbing their ancestors. There were a number of early species and geographical or regional variants of the same genus living at Saldauha Lagoon, Langebaan, South Africa.

Early hominids feared the unknown in their environment. Predators created stress. Having no defense, expressing their emotion and stress, they imagined spirits lived in their cave dwellings. Many transformed and transitioned as hominids. Growth of the early hominids was 20-30 years, four times faster than Modern human growth.

Kenyanthropus platyops evolved and may have lived 4-3.6-3 mya. Only teeth have been discovered. Discovery of a child's lower jaw with milk molar attached, were absent of dagger-like canines, unlike its ancestors who sharpened them by the lower teeth.

The early **Hominids** were Great Apes. Over time, in the above areas, other different hominid species were discovered. Small 3 foot tall primitive bipedal variants diverged with small brains and were specialized **Australopithecus (A.)** 4 mya. These variant groups beget from **A. anamensis** specie variants. Most of the **Australopithecus**, "near man", nicknamed "Zinj", (**Zinjanthropus**) lived in and along the East Africa Great Rift System in the Olduvai Gorge, except for A. robustus living in dolomite limestone caves and on the open savanna in South Africa.

From an outcrop ledge of the western margin of the Olduvai Gorge rift, in a ravine, hand bones were discovered in the Awash River Basin, Hadar, Ethiopia. The backs of the metacarpal had no ridges. These apes did not walk on their knuckles. The thumb could rotate making it possible to manipulate tools 2.52 mya.

Erect posture evolved from gene mutation miscreant as "markers". These changes in the anatomy were adaptations from their environment. Primative bipedal and palm walking was an adaptation carried on from the Common ape. The bipedal hominid habitually walks upright on two legs. The skeletal adaptation to this mode of locomotion and posture includes long legs stronger than the arms. The hand's muscular thumbs are opposed and rotate to touch any finger for increased dexterity. The feet are uniquely adaptable for bipedal walking, pliable platform shock absorbers, with an arch. To be bipedal a joint of the foot bone medial cuneiform articulates with the base of the big toe, which orients the big toe to line up with the other toes providing a strong toe push off for effective bipedal stride in all hominids and humans (man). The muscular buttocks and thighs may have permitted sprinting and long distance walking. Changes from the ridged back to a spine curve in the neck and lower lumbar vertebrae placed the center of gravity for the body in the pelvis area. The hole in the spine center accommodates nerves allowing for better breathing.

Australopithecus afarensis (AFA) variant has the same relationship and form as an early version of ANA. AFA beget from A. anamensis. For 900 thousand years the species was unchanged. AFA was known as the, "**Laetolil Hominid**". It lived in and along the East Africa Great Rift System, west and east margins 3.9-3 mya. A jaw and other remains were discovered at the Maka site in 3.4 million year old stratum of the Awash River, Afar Basin of the flood plain ridges between two streams, north of Adgantoli and east of the Bouri-Modaitar (Modaj), Ethiopia. Their fossils look like those discovered of AFA in Ethiopia 2.5 mya.

A female AFA variant fossil named Lucy, the southern ape Denkenesh, from the Ethiopian Afar language meaning, "you are wonderful", evolved 3.18-2.9 mya. She was discovered at Hadar, Ethiopia, Africa. The discovery of this female fossil had the anatomy of an ape and some adaptive traits toward becoming a early near Hominid.

Her features: the skull; brain was distinctly smaller like a chimpanzee, primitive teeth with male canines, massive jaw muscles, forward extended muzzle, projected snout, large brow ridge, the semicircular canals of the inner ear were not developed for the sense of balance needed to be capable of biped fast walking or running on two legs without falling. However, she had large ape pelvic iliac blades for the attachment of the upper thighs and the major abductor muscles that move the upper leg upright for biped walking. The angle of the **thigh bone** and the flattened surface at the **knee joint** end is different than quadruped apes, suggesting she walked upright on two legs. Her muscular arrangement of her pelvis, hips, and buttocks made it difficult to climb trees. She walked with a waddle and a shuffle. She may have walked with a slight bent leg motion, with her shoulders upward tilting, and with a free striding gait. Her torso; short neck, flat chest, ribcage flared out, and large muscular abdomen. Her pelvis was identified with the crescent indention on the inner edge of her pelvis that indicates her sex and a large sacrum (the series of tail bones) restricted the size of the birth canal. She had short legs, long arms with the upper arm humerus bone facing upward as in apes. The fingers and toes curved for typical tree climbing and vertical suspension. The male was larger than the female. Males were 4 feet (f) 8 inches (i), 99 pounds (p) and the female 4 f, 64 p, she was 3f 6i to 4f tall.

Other A. fossils were discovered northwest of Hadar, Dikika, Gona, Lake Tana, Yardi, and East Afar depression of the Awash River, south at Boda and Aramis, Ethiopia, Africa.

Australopithecus africanus (AFR1) variant lived in the Tanzanian Great Rift Olduvai Gorge 3.7-2.9 mya. Beget from ANA variant. Late **AFR1** was going through a transformation toward Archaic A. habilis 2.9-2.4-1.6 mya.

AFR1 features: evolved with a spinal mutation trait; a hole through the spine where it left the brain, it lay at the base of the skull, not toward the rear, as it was for other ape quadrupeds, and the spine was curved not straight like other apes 2.5 mya. This evolved spine adaptation was from their environment, and is a link to, but, is not even close to near apeman. There were other species variants that evolved with other man features. Males were 4f 6i 90p and the females 3f 9i 66p. AFR1 and AFR2 features: They had a robust powerful upper body build, small cranium, wide cheek bones, boney brow ridge, large molars, and jaw muscles. Flared iliac bones of the upper pelvis allowed upright walking without lurching from side to side. The lower pelvis was ape like to accommodate the powerful leg muscles used for climbing. They were bowlegged, with the legs and feet directly under the pelvis enabled the articulation for two legged bipedal movement and mobility on the ground 2 mya.

The Sadiman volcano erupted many times blanketing soft ash on the savanna at the edge of the Serengeti and Laetoli Plains 3.75-3.35 mya.

Thirty miles south of the Olduvai Gorge two sets of trace **fossilized foot prints** were discovered in a dry river bed at Laetoli in Tanzania, Africa 3.6-3.56 mya. A. africanus AFR1 was the only ape living in that area that could have walked upright, bipedal, and capable of striding between steps at that time. These footprints were devoid of the splayed big toes. The foot prints of the two, were side by side, in the once rain wet volcanic ash. The 45 feet long series of closely parallel consecutive in step footprints were made by two primitive bipedal apes. One set was smaller than the other with the big toes almost parallel to the other four toes and the axis of the foot. The footprints show a sound strike with the heel followed by a push-off with the big toe to propel the body forward. AFR1 had the unique ability of walking with their hands free, with the possibilities to gather and carry. This new freedom of the arms posed challenges.

The other **A. africanus 2** (AFR2) variant lived 3.3-2.1 and 3-2.3 mya. Local tribesman referred to them as **"Zinj"**, **(Zinjanthropus)** meaning "the apeman", living in the area around Taung. Makapansgat, and Sterkfontein, west of Johannesburg, South Africa. Where AFR2 beget has not been clearified.

The small skull of a AFR2 juvenile a little bigger than a grown man's fist was known as the "Taung child". This fossil was removed from a limestone quarry at Taung, South Africa, southwest of Johannesburg, in an area known as Bechuanaland 2.5 mya. AFR2 was going through a transformation toward Archaic A. habilis 2.

Australopithecus bahrelghazali (BAH) variant fossil of a jaw with teeth was discovered in Chad, Bahr el Ghazal, Africa, in the Djurab Desert of an ancient lake shoreline, 1500 miles northwest of Hadar, Ethiopia and may have lived between 3.6-3 mya. The features are unknown. No other lineage has been established.

Australopithecus gracile (GRA) variant had light features and large jaws. GRA lived 3.18 mya, at Saldauha Lagoon, Langebaan, and Hoedjies Punt, northwest of Cape Town, South Africa. GRA beget from AFR2 variant.

Australopithecus garhi (GAR) variant meaning "surprise" in the Afar language, lived in the Middle Awash open plain territory of the Afar clan Alisera in Ethiopia 2.5-2.3 mya.

GAR features: skull; brain was 450 cubic centimeters (cc), small front teeth, molars and premolars. This clever long legged hominid eking out a furtive stealthy existence among larger and faster predators avoided their jaws. It had traits similar to Archaic H. erectus.

Australopithecus Naledi (NAL) evolved in a place in time possibly 2.5-2 mya. "Naledi", means star in the Sotho language. The Rising Star site is 25 miles northwest from Johannesburg, South Africa, at a limestone dolomite cave system with cascading white flowstone vails, leading to the remote Dinaledi (small) chamber, where a lower jaw with teeth intact was observed. Further observations discovered the bones were not incased in stone (fossilized) in the cave sediment. The jaw was removed and determined to be hominid. Many bones were excavated to make a composite skeleton of a hominid apeman.

NAL features: modern adaptive shape skull (more rounded); small brain estimated to be 560 cc male and 465 for female, round shaped chin, teeth had small molars and cusps, spine was curved, shoulders were positioned for climbing and hanging. NAL short flared pelvis was not like modern adaptations. It was longer in the front and rear. It had long arms; hands with palm wrist and thumb having advanced traits, and long curved fingers. It had long slender legs; with attachment for strong muscles used in advanced traits for bipedal gait, except the feet had slightly curved toes adapted traits typically used for long distance striding. NAL males were 5 feet tall and weighed 100 pounds.

The 2nd Indeterminate portion of a skull was discovered in the east margin of the rift, south end of Lake Baringo, at E. Chemeron in Tanzania/E. Lake Victoria, Kenya in stratum dated to 2.3 mya. Specie analysis was not determined.

The 3rd Indeterminate jaw was discovered at Hadar, Ethiopia. Scientific opinion determined the species was between Arr and A. 2.3 mya.

Australopithecus sediba (SED) variant evolved adapted to their environment 2.2-1.9-1.78 mya in Sterkfontein's red rock wall limestone caves, 25 miles northwest of Johannesburg, South Africa. At an eroded limestone cave called Malopa (death trap) fossils of SED were discovered showing signs of transforming toward early H. erectus (EE) 1.99 mya. SED beget from AFR2/early erectus EE variant. SED lived with cousin Australopithecus NAL.

SED features: skull; larger brain with a frontal brain reorganized. Big brains are useful and require higher calorie foods, small jaws and increasingly smaller back teeth with chewing muscles, primitive molar cusps and cheekbones relating to earlier Australopithecus, and a snoot projected nose. Their bodies were lean developed, narrow human like pelvis built for bipedal stride, less body hair and may have developed sweet glans. They had long legs, modern ankles, primitive heal bones, and flat griping hands. They ate plants and tubers from under the ground. Meat was scavenged from lion kills and other small animal meat kills. Animal bones were crushed for the marrow.

A community of variant apes lived west of Cape Town, South Africa, near the Atlantic Ocean, at Saldanha Lagoon northern margins at Hoedijies, Langebaan, and Geelbek.

The Cradle of Humankind of Gauiteng Province

Limestone formed in the Atlantic Ocean 2.3 bya. Over millions of years after the ocean retreated acidic ground water dissolved out the dolomite from the breccia limestone forming the calcium carbonate water that formed the caverns. The Province encompasses the Johannesburg, South Africa area. In the lower caves lime breccia brine drippings calcified the fossils. Bone tools were used for digging 1.8-1 mya.

In a cave at Swartkrans, South Africa, hand bones were discovered in a lower cave at Drimolen, 25 miles northwest of Johannesburg, South Africa. Thousands of fossils were discovered at Sterkfontein, near Johannesburg in stratum dated to 1.5 mya, Burton near Taung (Taung is near Sterkfontein near Joannesburg), and others in the northwest Witwatergerg, Kromdraii, and Swartkrans. Another site outside the province was Krugersdorp.

The Family of Hominid Australopithecus
Paranthropus 2.8 to 1 Million Years

There were three species with **Paranthropus** skull traits transformed into two groups having special eating adaptations. The evidence indicates they were the first of these species to diverse. These variants evolved as specialized eaters adapted to tough tubers and other hard foods found in their environment. They were known as "Near Man", and were 7/8 ape and 1/8 man or apeman.

Their features: thick muscled body; skull; sloped back into the skull roof with a sharp center bone crest keel (sagittal crest) running from the back to front serving as the attachment for a sheet of massive chewing jaw muscles known as **paranthropus** attached to their large face, large jaws, large teeth; with large molars and small canines.

Some knowledgeable authority suggested AETP and BOIP are the same species. There is no fossil evidence they ever lived together 2.8-1.2 mya. AETP mingled with BOIP, because of later discoveries of fossils of AETP/BOIP variants in 2.6 mya stratum.

These Australopithecus did migrate and adapt to their changing environment, except for the surviving few, all parantropus were extinct by 1 mya. They used primitive stone tools to scavange and process animals 1.9-1.2 mya.

Australopithecus aethiopicus paranthropus (AETP) lived 2.8-2.6-2.2 mya in Omo and West Lake Turkana near the Ethiopian and Kenyan border at Lake Turkana and Lake Rudolf, Kenya, Africa. AETP and BOIP variants were living in Ethiopia, Kenya, and the southern part of the Olduvai Gorge Rift in Tanzania 2 mya. AETP beget from Australopithecus afarensis. AETP mingled with BOIP variants 2.6 mya.

AETP features: body was muscularly developed, Skull; small brain, an enormous brow ridges, and a large flat robust face. Most of them were extinct because of food resource changes by 2.2 mya. Males were 4f 6i 108p, female 4f 1i 75p.

Australopithecus boisei (BOIP), "Zinj", zinjanthropus paranthropus variants known as the "East African Man", 2.7-2.2-1.2-1 mya, lived in East Africa, along the Rift System margins at Omo, W. Lake Turkana within the caves of the Olduvai Gorge, Kenya and Peninj, Chesowanja, and Koobi Fora, Tanzania 1.75 mya. BOIP were isolated 500 miles south of Laetoli, at Malema, northwest margin of Lake Malawi, in Tanzania. Other locations were the Shungura Formation at Konso, Ethiopia.

BOIP features: Skull; face resembles AETP, small brain, jaws more rounded in front than those discovered at Hadar 1.6 mya. The teeth front and back were more evenly proportioned than other groups of Australopithecus.

Australopithecus robustus paranthropus (ROBP) variant "Beside Man" or "Near Man", or "Handyman with skills", were the first group of this kind that evolved at Saldauha Lagoon, Langebaan, northwest of Cape Town, South Africa 2-1.2 mya. Also, a second group living in Drimolen, 25 miles northwest of Johannesburg, and 4 miles north of Sterkfontein and Swartkrans South Africa 2-1.2 mya. This area is in The Cradle of Humankind where they lived in the open not in caves. ROBP fossils were discovered at the Taung site, 1 mile northeast of Sterkfontein and at Kromdraii, South Africa.

The first with a larger brain were the **Australopithecus robustus paranthropus (ROBP)**, living and adapting to the arid and open land at Taung, Stirkfontein and Makapansgat, South Africa 2 mya. They discovered other foods were seasonal and grew in patch areas. In the wet season, meat was abundant and many supplemented their diets as meat eaters. They had greater cooperative participation and longer hunts. Living in larger groups, they ranged over larger areas. ROBP needed some kind of communication with each other to be more efficient. The apes only used tools when it was worthwhile.

ROBP features: small stature Hominids were robust, strong, and stout. The skull; head sloped back into the skull roof with a sharp center bony keel crest, females did not have a crest on their skull, large brain, large face because of their diverse large jaw muscles for crushing hard plant food, sunken nasal areas on their flat faced, heavy enameled large teeth with large molars, small front incisors and canines lacking the honed edges, a weaker chin, protruding brow ridges, and large eye sockets. Males were 1/3 larger 4f 4i 88p than the females 3f 7i 71p.

Animal bones were found that show heavy wear on the tips and parallel along the shaft. These tools were use for digging and extracting plant roots. ROBP hands thumbs had different muscular attachment sites indicating the flexor pollex longus muscles were necessary to allow for a precision griping with the fingers. The muscular attachments would be necessary in the advancement of the use of tools. They spent considerable time squatting on the ground. The bipedal stance was not an advantage. ROBP adapted to the dry seasons living on tough food, roots, tubers and seeds. They were highly adaptable until the climate got cooler and drier 1.7 mya when the woodlands turned into grassland and prairie. Most of them did not migrate and were extinct by 1.3 mya, due to resources and its inability to adapt to its changing environment.

ROBP variant fossils were discovered in 500-400 thousand year old stratum ca. The fossils eroded up from lower sediment 2 million years old. Creditability is lacking and the needed facts to prove the claim.

THE FAMILY OF HOMO
3.9 MYA TO 8 TYA

Homo (H.) is the genus of primate including modern man comprising of several extinct A. Hominid species and man, known as, "those who walked erect".

Human is a being having the form or the characteristics of a person; man, woman, or child.

Large carnivores prowled Europe; Saber Toothed Cat and Homotherium lived 3 mya to 500 tya. Hyenas and Pachycrocutq lived 1.5 mya to 500 tya, when they all became extinct.

There are hominid Australopithecus (A.) gaps in the fossil record events of time lacking the facts proving when the first true Homo evolved. The Scientific Community needed a transition threshold to H. Not having any answer or any proff, they argued in a concenses for the arbitrary concept in the quest to determine the lost link between A. and H. New A. to H. opinion was based on A. brain size development larger than 600 milliliter (ml), 1ml=1/1000 liter. This was subjective reasoning, and irrational action in absents of facts. It did not take in consideration other factors such as body mass or the difference between male and female forms. A more positive bridge to the Primate gap in the fossil record is to resolve a discovery of A. hominid fossil with ape traits transitioned with enough modern man adaptive trait qualifiers to be classified H. genus. Creditability is lacking. We need the facts proving the claim. There may be other physical features that were not taken into consideration in the classification; Child, woman, body size, and age are factors in question with big brain changes. This ape with the brain size to qualify as Homo was established to have lived 1.8 mya by the same individuals establishing brain capacity. This first brain transition fossil has not been discovered that fits the subjective analysis.

Arbitrary Concept of the Brain size difference and classification was established to transition from A. ape brain size to ape Homo genus by Phillip Tobias, John Napier and Louis Leakey in milliliters (ml), 1 ml=1/1000 liter. Some of the Scientific Community has accepted these opinions contrary to any valid facts. Ape small vs ape H. large brains responsible for the determination of that event they established to be 1.8 mya. Their bias was based on A. ape erect posture, bipedal two leg gait of a quadruped, precision handgrip, and having a brain size containing originally 600 ml capacity was changed to 600 cc or 19.25 ounces or larger for A. hominids to become Homo genus. One liter = 1000 ml; one ounce equals 31.25 ml; 1 pint, 2 cups, 16 ounces = 500 ml, 5 cubic meters (cm) was converted to 625 cc, 6 cm=1296 cc, 6.25 cm = 1526 cc, 6.5 cm = 1785 cc, 6.75 cm = 2076 cc, and 7 cm = 2401 cc, and 61 cubic inches = 1000 cc. The first of the **Homo genus** redefined Australopithecus ape classified brain size units of capacity was 675-680 ml or 600 cc.

Homo was more advanced 1.8 mya. Many were nomad variants. They migrated north along East Africa into Asia and Europe 1.6 - 1 mya to 700 tya. They were sole survivors of small groups of **Archaic Homo sapiens (AS)**, an elite sapiens species 800-600 tya, and 500-250 tya.

Between the periods of 350-120 tya few fossil remains were discovered in Europe.

Europe was cold, few ventured from Africa primarily because they needed to adapt to the cold climate and a new way to survive. The Homo adapted to the changes in cooler and drier climate when the woodlands turned into grassland and prairie. Archaic sapiens were scavengers, meat eaters, and had primitive tools for hunting. There were groups of manape and man who were cold weather adapted 200-100 tya.

The Magantereon White Saber Tooth Cat was prowling in Spain 1.2 mya. Not an environment where scavenging Hominids or AS with primitive tools would want to live.

HOMO ANCESTRY

Archaic Homo's were commonly referred to as; A. and H. habilis, H. ergaster, A. and H. rudolfensis evolving 3.9-2.5 mya, and H. erectus 2.4 mya to 30 tya.

Their features: skull; jutting jaws, a heavy brow, flared cheeks, flat wide nose, muzzle like mouth, and canine teeth with large roots smaller than their cousin's 2.5 mya. The males were 5f 100p, female 4f 60p. They coexisted with AH, RUD, AETP, ABOIP, AROBP, AH/EH, AEG/EEG, AE/EE variants. Archaic Homo's were transforming toward EE variant, resembling AEG. Time passed, Homo EE variants transformed as apeman three quarters ape and one quarter man. Archaic sapiens transformed as manape three quarters man and one quarter ape, and Modern H. sapiens were all man. Som of the AEG migrated into Europe transitioning as H. antecessor, who transitioned to H. neanderthalensis (NEA), all man, and was a close cousin of Modern sapiens.

Olduvai Stone Age Technology tool origin was in the Awash Valley System 2.6 mya. They were the oldest; simple pebbles, hammer stones, crudely fashioned choppers, and primitive flake stones with sharp edges. The tools were discovered at Omo, Hadar/Gona, Senga, Ethiopia, and Lokalei, Kenya, Africa dated to 2.6-2.3 mya, and were still used 1.83-1.53 mya. Tools discovered in river beds were used as crude cutting and chopping tools, littered about at large butchery site area among large animal bones. Use of these tools took a relatively high order of knowledge and dexterity to accomplish the butchering tasks. Along the border between Ethiopia and Kenya cleavers, scrapers, and crushing stone tools were discovered dated to 1.82-1.53 mya. Meat scavenging and animal kills gave Homo a new way to survive and provided the extra energy needed to grow larger brains.

Arachic A. habilis 1 (AH1) was transformed from AFR1 living 2.9-2.4-1.6 mya. East/North African Koobi Fora fossils were discovered 30 miles south of Olduvai Gorge in Tanzania, and at Lake Turkana (later Lake Rudolf) in north Kanya. They evolved with an average brain size of 510 ml. BOIP, AETP, EEG, ARUD and EH co-existed in this area.

Early H. habilis 1 (EH1) transitioned from AH1 variant and lived 2.3-1.4 known as "man with skills", "able or handyman".

EH1 features: skull; larger brain 675-680 ml or 600 cc, skull base was positioned above the backbone allowing for a spinal cord (foramen magnum), face, cheek, and teeth were smaller and narrower, large body; light boned, short legs lacking the hip joint locking mechanism necessary for bipedal walking, feet; big toes aligned parallel with the other toes, ankles turned the feet inward like an ape with a pigeon toed stance, no arch (flat feet), long arms with stout thumbs and a precision grip dexterous enough to have used stone tools found near the fossils. They were eating some meat protein scavenged from predator kills, but were not active hunters.

H2/A. robustus (HROB) variant fossils were discovered in South Africa 1.6 mya. EH2 mingled with A. robustus and others who were more populated there. They did not migrate out of Africa and became extinct.

H. habilis 2 (H2) lived 1.4 mya-800 tya. LH was transforming toward EE, with erectus traits that looked similar to ADEG 800 tya.

EH2 teeth, two arms, and a jaw were discovered. Also, numerous fossils of **A. robustus** were among the discovery with brain size of 630 cc. West of Cape Town, South Africa, at the Atlantic Ocean Saldanha Bay Lagoon, at the southwest margin, fossils of many A. species variants were discovered at Geelbek and Langebaan.

Over time the climate became drier, lakes became alkaline from rivers of dissolved salts and volcanic ash deposits, giving way to larger grassland.

Later **Acheulean Stone Age technology** tools were used; bifacial choppers with flukes on two sides with sharp edges providing a more efficient ax like chopper, scrappers, pointed awls, and disc shaped polyhedral tools dated to 1.5 mya-100 tya. Some of the bones from animals preserved in the stratum displayed cut marks from the tools.

H. georgicus (GEO) EH/early erectus EE variants having the primitive features of AH and advanced features associated with EE were living within small groups of individual hominids who lived together. In Asia, Republic of Georgia in lower south Central Russia, under ruins of the village of Dmanisi, built in the 9th century, a fossil of a lower jaw from a skull of EE was dated to 1.8-1.77-1.6-1 mya. Also, EH/EE were discovered of three adults and a teenager. One fossil of a 40 year old man had lost all of his teeth before his death. His brain was estimated to be 780 cc.

Archaic A. ergaster (AEG) variant evolved 2.9-1.5-1.8 mya, beget from A. afarensis.

Plesanthropus trensvaalensis (PT) living in Beijing, China were determined to be AEG/AE archaic erectus variant dated to 2.9-2.5 mya.

The young AEG male skull known as, "Nariokotome Boy", was discovered at Nariokotome, Kenya, Africa dating to 1.9-1.5 mya.

H. Arachic Pithecanthropus (API) variant and AEG/AE variant are similarly looking. These Asian Chinese may have evolved 2.4 mya. H. Pithecanthropus PI origin was difficult to establish the facts of the finds, because of Chinese influence and conflicting testimony. API variant transformation of PT AEG/AE variant may be the same 2-1.7 mya.

AEG of Africa or Archaic A. heidelbergensis (AHEB) of Europe may be the same specie, if AHEB ever existed.

Early H. ergaster (EEG) variant transitioned from AEG 1.8-1 mya. Known as, "Koobi Flor" meaning "Workman", by association to stone tools discovered in East Africa at Lake Turkana, Kenya. EEG fossils were discovered in southern India 1.6 mya. EEG was more advanced and looked like modern humans, especially in the way they walked. EEG was transforming toward early erectus EE. EEG variants migrated from Africa to eastern European Russia, and western Asia 1.1-1 mya. Some researchers regard the smaller EE and EEG are the same specie variants. However, there were many specie variants as they mingled with others 1.8 mya.

Both EEG and EE had the same features: thick skull; large brain estimated to be 871-984-1097 cc or 1.85 liter, robust face, bony brow ridges, with traces of a bony keel (sagittal crest), large chinless lower jaw, small teeth, forward protruding nose not like Early Homo, no forehead frontal lobe development for complex thinking. They had a narrow pelvis needed for a more efficient bipedal, slim build with a narrow trunk to minimize heat absorption and for more efficient cooling, a primitive mix of advanced ribcage, waist, backbone with a narrow passage way for the spinal cord and nerve control for the ribcage to allow for improvements in walking, breathing, and vocal emissions all articulated with the pelvis. They had long modern arms and legs and were 6 f tall.

EEG variant was discovered at Atapuerca, Spain 1.1 mya (900 tya). The Spanish scientists discovered EEG variant and classified their subjective controversy based on the H. heidelbergensis HEB 600 tya jawbone resembling the traits closely to EEG. These dates are in conflict and may be 780/710/575 tya when some EEG transitioned toward H. antecessor (ANT) variant and then transitioned as H. neanderthalensis (NEA) Modern Man's cousin.

Advanced H. ergaster (ADEG) transitioned from EEG, evolving 800-250 tya. Groups of ADEG migrated out of North Africa eastward settling in various areas of Eurasia, Middle East and Asia and and northern China. They became cold climate adapted. They migrated from China back to Eurasia, Middle East, Europe, and along the Mediterranean Sea south shoreline in Africa 250 tya.

H. rhodesiensis (RHO) ADEG variant lived 300-125 tya. RHO fossils were discovered at Broken Hill, now the Kabwe cave system in Northern Rhodesia, now Zambia, Africa, in surrounding rock stratum, analyzed by relative dating to be 800 tya. RHO fossils were discovered in a limestone sulphide ore mine cave 89 feet below the surface. Tom Zwigelar discovered the fossils and Arthur Smith Woodward in London, England named this specie RHO, not AHEB.

RHO features are similar to modern man and ADEG. The skull; brain was 1300 ml, thick bone and robust to support the neck muscles, large prominent horizontal ridge at the back of the skull, low sloping forehead, and heavy wear on the teeth indicating a diet of hard abrasive plants, tubers, and roots. RHO used Acheulean Stone Age tool technology; to fabricate stone blades and axes sharpened on both sides, cleavers, and stone spear flakes used as hammers to cut ivory, bone, and antlers, or for digging roots and tubers. These kinds of tools were used until 250 tya.

NOTE: RHO variant with ADEG and EEG features is the subject of a controversy concerning the existence of archaic A. heidelbergensis (AHEB) being RHO, but, has not been proven. There is no proff AHEB ever existed. No AHEB influencing relationship could have existed to RHO due to differences in discovery age. See Pre H. Neanderthalensis Controversy A/H. heidelbergensis.

Archaic H. rudolfensis (ARUD) variant transformed from AH1 and lived in the Olduvai Gorge at Lake Rudolf, in Tanzania, Africa, 2.8 (2.5-1.9) mya. Males were 4f 4i 82p, females 3f 10i 71p. Some ARUD variants were transforming toward early habilis EH 2.1-1.5 mya. Some researchers suggest H. rudolfensis and archaic habilis AH1 from Tanzania are the same specie. There were some ARUD variants who transformed toward Early H. habilis EH. ARUD was begat from africansis AFR1.

At Lake Turkana, Kenya, Africa a fossil of ARUD variant was discovered with a cranium exhibiting a large brain cavity of 800 cc, more rounded than AH, with small teeth, long broad face, and flatter brow ridges. There were many early variants that had developed large brains. This skull lacked the protruding brow ridges. It may of had a relatively high order of knowledge and dexterity. Simple Olduvai Stone Age tools for cutting and chopping were discovered dating to 2.6 mya ca.

A. kenanthecus rudolfensis (KERUD) morphic variant was a surprise to scientific experts. KERUD fossil is dated to 2.3 mya. It was a little more than half early Hominid ARUD and Cercopith ecidbou. DNA difference did not match. The body part was a fact. The monkey part of the body could not be a fact due to differences in genera.

Archaic H. erectus (AE) transformed from A. afarensis and was an apeman who evolved in Ethiopia, Africa 2.5 mya. AE were known as, "true man". As the environment became arid they foraged in more open grassland 2 mya. Brains do not tolerate heat and bigger brains need more cooling. They lost body hair and maybe developed sweat glands providing cooling for the body. **Plesanthropus trensvaalensis** (PT) living in Beijing, China and later was determined to be an AEG/AE variant dating to 2.9-2.5 mya.

Arachic H. Pithecanthropus (API) variant was Chinese. It was influenced by the culture of AEG/AE variants and may be the same species. API origin was difficult to establish because of Chinese conflicting assessment.

Early H. erectus (EE) variant transitioned from AE. Groups of EE lived in East Africa, at Lake Turkarna, Kenya 2.1-1.9-1.8 (1.4) mya. Fossils were discovered in southern India in stratum dated to 1.8 mya. No fossils of EE have been discovered in Europe. EE did not migrate into Europe. Some researchers regard the smaller EE and early egaster EEG as same species. Many specie variants mingled with others 1.8 mya. There were many AE and EE variant forms by 1.8-1 mya.

Both EE and EEG had the same features: thick skull; large brain estimated to be 871-984-1097 cc or 1.85 liter, robust face, bony brow ridges, with traces of a bony keel (sagittal crest), large chinless lower jaw, small teeth, forward protruding nose not like Early Homo, no forehead frontal lobe development for complex thinking. They had a narrow pelvis for a more efficient bipedal, slim build with a narrow trunk to minimize heat absorption for more efficient cooling, a primitive mix of advanced ribcage, waist, backbone with a narrow passage way for the spinal cord and nerve control for the ribcage for allowing walking, breathing, and vocal emissions all articulated with the pelvis. It had long modern arms and legs and was 6 f tall.

At Sangeran Basin, Java, Indonesia, a Southeast Asia Javanese fossil of EE with a larger brain was discovered dating to 1.6-1 mya, known as "Java Man". Other fossils were discovered at Sumatra, Sangeran and Samgung. EE fossils were discovered at Ngangdong, Sambungmacan, and Ngawi, Indonesia 300-30 tya.

EE (**sinanthropus pekinensis and soloensis**) variants, "Peking Man", were discovered in Beijing (Peking), China at the Dragon bone hill site in the valley of the Zhoukoudian caves 31 miles southwest of Beijing dated to 420 tya, in stratum dated to 800-600 tya ca. The pumice is older than the skull. The skull from upper land slopes, eroded down into the older stratum deposit. Also, discovered were stone tools and signs of the use of fire. **Sinanthropus (Sinpithropus) pekinensis** from Zhoukoudian, China is now known to be EE.

Some researchers regard EE and EEG as same species. EEG was transforming toward EE. EE/LH (GEO) ADE/ADPI variants were nomads moved on to Sumbawa Island, and then on to the Flores Chain of Islands where they became isolated from further migration to the FLO culture.

Late H. erectus (LE) transitioned from EE having the traits of a manape from the neck down, close to Arachic H. sapiens (AS) 825-800-250, and Early H. sapiens (ES) 800 tya. In the Awash region, at Bodo, Ethiopia, Africa stone hand axes and cleavers of the Acheulean type were discovered 800-600 tya.

LE features: skull; larger brain 750-1100 cc with a brain case low and heavy, slight crest bone from the forehead back along the center of the skull, smaller chewing and neck muscles (carry over from the Paranthropus hominid ancestry), protruding brow ridges known as, "beetle brow" that shadowed their face, a flat wide nose, a muzzle like mouth with a <u>full chin</u> lower jaw, and ears set back on the head, their throat voice box, (hyoid bone) was developed enough for rudimentary language, probably made meaningful utterances, but no linking of sentences. LE had a large body for greater mobility. They were tall with long legs and forearm more ape like.

LE de-fleshed humans after death and their body parts were cannibalized. There were cut marks on the skulls. They migrated out of Africa into Asia, and took their tools and knowledge with them after 700 tya.

China-Australian Theory: time passed 900-186 tya some of the LE/ADE, AS/ES/EMS/PI and other nomads from northern China may have migrated to New Guinea, when shorelines were 400-270 feet below sea level. How they migrated and navigated has not been verified, however, Homo fossil discoveries verify they lived in New Guinea. The seamount and coral reef islands of the Torres Strait land bridges provided crossing at Cape York Peninsula, Australia 75-70 tya.

Fossils of LE/EPI variants from Yunxian, Chinese province of Hubei, were discovered dating to 600 tya. LE/EPI features: skull; small rounded head not angled back, it had a bone crest, high cheekbones, with straight brow ridges. Their body was slender like EEG.

Advanced H. erectus (ADE) variant transitioned from LE, lived 300-260-30-28 tya. ADE was transitioning to Early H. sapiens (ES) and Early Modern sapiens (EMS) as manape. Also, EMS was transitioning to Late Modern sapiens (LMS) all man. Fossils have been discovered in China at Dali, Maba, Jinniushan, and Xujiayo, in Skhul and Qafzeh Israel 200 tya, in Omo, Kibish, Herto, Ethiopia 195-155 tya, and in the Middle Awash Valley west margin of the rift at Herto, Ethiopia 160-145 tya. ADE and others migrated out of Africa 100 tya into Europe and Asia. The males were 5f 10i 139p and the females were 5f 3i 117p. Over the long existence of H. erectus it coexisted with ARUD, BOIP, ROBP, AETP, EH, EEG, NEA, API, AS, ES, EMS, LMS, and AMS.

ADE developed Acheulean Stone Age tool technology with the knowledge of mentally mapping the strike "knapping" a black basalt core rock at the correct points to produce sharp flacks on all sides. They crafted tear drop shaped spear heads using the technology. They crafted symmetrical tri-side **hand stone axes** to fit snuggly in the palm of the hand. The way it was held could cut, crush, or batter, animal skin and bones. Bones of antelope, horses, and other animals exhibited tell tale cut marks on the surfaces from stone tools. In the butchery other bones were discovered of a single ADE hominid with adapted human traits; lower jaw, femur, thighbone, upper leg, and an arm. They painted themselves with ocher, participated in danced rituals, and were fire makers.

Java Trench Theory: LMS variants may have discovered a way to successively navigate the Java Trench from Bali, to Lombok a steamy volcanic island. Who these humans were has not been verified. The conflicting dated stone tools were fabricated from volcanic Lombok basalt lava dated to 840-700 tya and was used by humans on the islands of Sumbawa and Flores 95 tya. It does not tell when tools were fabricated from the volcanic basalt formation. Stone tools from the Lombok lava were discovered on Flores Island in rock layers and were dated to 700 tya.

NOTE: In the 1860's, Alfred Russel Wallace, established a theory about the migration of man and animals not crossing the Java Trench between Bali and Lombok Island. An invisible biological barrier, known as the "Wallace Line", is located at the center of the Trench, between Bali, Indonesia, and Lombok Island. The 15 mile crossing of the Strait of Java Sea Trench prevented the Asians from navigating treacherous currents flowing northeast from the Indian Ocean into the Java Sea.

H. floresiensis (FLO) variants had traits from early H. erectus EE/LH late habilis (GEO georgicus from Dmanisi, Republic of Georgia). These nomadic EE/LH (GEO) ADE/ADPI variants migrated south to Sumbawa Island, and then on to the Flores Chain of Islands where they became isolated at Soa Basin, Liang Bua Cave, Indonesia, in the Manggarai language known as, "cool cave". They transformed to H. floresiensis FLO, living the Hobbit of FLO culture 95-18-12 tya. FLO fossils were discovered in Liarg Bua Cave. Over thousands of years they evolved adapted to their environment by dwarfing in size and strength. It was less important when isolated, having no predators to be concerned about. They were hunters that killed with hand held wooden spears and butchered the kill with basalt stone tools dated to 840-750-700 tya ca. When the tools were fabricated is unknown.

FLO small features: Skull; brains were small 17 tya and were 710 ml (1/3 of modern man) by 13 tya. The frontal and temporal lobes had developed consistent with higher mental processing, planning and reasoning, aft sloping forehead, nutcracker jaw, and bony arched brow ridges. It had a long 8 inch foot resulting in a high stepping gate with short legs. They were 3 to 4f 9i to 5f 5i tall, weighing from 88 to 110 pounds. FLO cooked in open hearths using fire.

Tools were discovered on Timor Island, south of the Flores Islands. No human fossils were discovered.

On Sumatra Island, Indonesia, Toba Volcano eruption creating a 63 mile wide crater forming Lake Toba. Human fossils were discovered.

In China, Sinchuan Province, near the Yangtez River, in a cave called Longgupo, a jaw with two teeth of an EE has been paleo magnetic dated at 2-1.8 mya. EE fossils discovered later were dated to 1 mya.

Early Pithecanthropus (EPI) EE variant transitioned from Archaic PI 1.7-1 mya, were living in Sangiran China, Southeast Asia, and in Mojokerto, Java, Indonesia. EPI was an apeman.

NOTE: Fossils were discovered by Eugene Dubois a Dutch military medical officer in the 1860's along the banks of the Sole River, near the village of Trinil, on the island of Java, Indonesia. Part of a skull cap and a tooth were dated to 1 mya. He named his discoveries **Anthropithecus** meaning man-ape and later he corrected it to apeman. He used the name already used for fossil Hominids from the Siwalik Hills, India. Dubois thought the name was less appropriate, so he turned it around as **Pithecanthropus**. He later discovered a left thigh of LPI/EE or LE/ADE variant which he added to his collection.

EPI/LE/LH/EPI variants were evolving 700 tya. There were EH/EE/LE/ADEG/API variants fossils discovered in China 700 tya.

Late PI (LPI) LE/ADE variant transitioned from EPI 1 mya-600 tya. Some of the LPI group migrated from China back to Eurasia, Middle East, Europe, and along the south shoreline of the Mediterranean Sea in Africa 600 tya. Fossils of EPI/LE variants from Yunxian, Chinese Province of Hubei were discovered 600 tya. The features: skull; small head was rounded not angled back, bone crest, high cheekbones, and it had a straight brow ridge. Others were slender like EEG.

Advanced H. pithecanthropus (ADPI) LE/ADE variant transitioned from LPI 250-100-90 tya. Beget by H. ADE/ADPI variant dated to 100-50 tya.

Jinniushan (JIN) ADE/LPI variant fossil of a woman was discovered on the border of China and North Korea, the most northeastern zone of human inhabitation during the Pleistocene Epoch 260 tya. She was 5 f 6i tall and weighed 173 p. She was cold climate adapted with short limbs.

H. soloensis (SOL) ADE/ADPI variant was discovered in Ngandong, Bali, Indonesia dating to 400 tya. SOL features: skull; brain size 870 to 1149 ml, skull roof was long, chinless jaw, front teeth were small except for the molars, low bony brow was narrow behind the eye sockets, thick ridge of bone; vertical rounded ridge of bone (crest) from the front to the back of the skull, and another ridge horizontally around the back of the skull.

Chart 2.
Phylogeny Of Primate Hominid Australopithecus
Ape/Homo 4.9-1 Mya

Primate Population Loc. **	Species Variant
	1 Early H. pithecanthropus (LPI)
	1.2 Archaic H. antecessor (ANT)*
	1.4 Late H. habilis (LH)
	1.6 Early H. habilis/A. robustus (EH/ROB)
	1.77 H. georgicus (GEO) (EH/EE)
IND GE IT SP ET	1.8 Early H. ergaster (EEG)*
	1.9 Archaic H. heidelbergensis (AHEB)*
	2 Early H. pithecanthrous (EPI)
SA	2 A. robustus paranthropus (ROBP)
CH ET	2.1 Early H. erectus (EE)*
SA	2.2 A. sediba (SED)
KN	2.3 A. kenyanthropus rudolfensis (KERUD)
2.3 Early H. habilis (EH)	
CH KN ET SU AL MO	2.5 Archaic H. erectus (AE)
ET	2.5 A. garhi (GAR)
SA	2.5 A. Naledi (NAL)
CH	2.5 A. pleasanthropus trensvaalensis (PT) China AEG/AE
CH	2.5 Archaic H. pithecanthropus (API) AEG/AE
ET	2.5 H. RUD cercopith ecidbou (monkey apeman) (RUDCE)
ETTA KN	2.7 A. boisei paranthropus (BOIP)
TA KN	2.8 A. aethiopius paranthropus (AETP)
TA KN	2.8 Archaic H. rudolfensis (ARUD)
TA ET KN	2.9 Archaic A. habilis (AH)
CH SP ZM/RHO	2.9 Archaic H. ergaster (AEG)
SA	3 A. africanus (AFR2) South Africa
	3.18 A. gracile (GRA)
CHA	3.6 A. bahrelghali (BAH)
KN TA	3.7 A. africanus (AFR1) North Africa
ET KN	3.9 A. afarensis (AFA)
ET	4 A. kenyanthropus platyops (KEP)

Hominidae

Time mya   4   3   2   1

* See text for Pre H. Neanderthalensis Controversy

** See Primate Population location of species chart 4 African, European, Asian

THE ELITE SAPIENS

THE FAMILY OF HOMO SAPIENS

Homo sapiens evolved as human having the form of a person. They were **Pagan** individuls with no religion.

Archaic H. sapiens (AS) ADEG/LE variant lived 825-800-250 tya. AS was half ape man and half man ape. AS transitioned to Early Homo sapiens (ES) manape. Neurological mutations of their brains mentally changed AS. They could think and reason, were cold climate adapted, and appeared in Europe 500 tya. In Heidelburg, Germany, a mandible lower jaw of ADEG, questionable EHEB NEA, AS variant was discovered in 500 tya stratum.

AS features: skull; face appeared to be a combination of ADEG LE it included a mid face projection like a beak throughout Europe, Eurasia, Middle East, Africa, and Asia. They were 5f 10i tall. It coexisted with European ENEA, but were not known to socialize with them 300 tya.

To survive in the winter months when food supplies shutdown, AS developed a complex social networks to plan and organize small groups that separated from others for extended periods in Europe.

AS populated Africa from Morroco, at the Atlantic Ocean and along the Mediterranean Sea south shoreline, east to Egypt, south to Ethiopia and the East Africa Great Rift System along the Indian Ocean to South Africa.

In Europe there were large butcheries and shared cooperative hunts of reindeer, ibex, and bison. AS learned cooperative hunting rather than scavenging. They crafted wooden spears sharpened on one end for hunting and had a rounded base on the other end used for thrusting to assist in kills and for digging for ground food. They learned how to thrust the spear into an animal to kill it. Stone hammers and flakes were used to scavenge meat and bone marrow. Later technology developed the throwing projected stone balanced spears. Speech and a language was developed making it useful for strategic planning and passing on information critical to cooperative hunting. Large cleaver stone flecks 7 ½ inches long and other stone tools were used to butcher large animal carcasses, scavenge meat, and bone marrow needed to feed high energy metabolically expensive bigger brains of Man.

Their dwelling structures were constructed like those of an African bushman. The foundation was a circle 9 to 13 feet in diameter lined with stone and bone. One elephant tusk was used as the center post. Hearths were observed scattered over the area inside and outside the shelters. Some type of cultural social group rituals took place in a large 27 foot shelter lined in a circle with smooth stones and bones. Inside was a large quartz anvil set between the horns from a large bison. Scattered around it were fractured human skulls.

An Mammoth elephant tibia was discovered inscribed with 7 regular lines in one direction and crossed at right angles by 21 other lines, all parallel to each other. These deliberate marks indicate abstract thinking and symbolic behavior had developed 400 tya.

Sixty miles from Schoningen, in Thuringia, in a Neumark Nord Coal Mine near Bilzingsleben, Germany (former East Germany) by a spring fed lake a discovery was made of a social culture that existed there 412-320 tya. Workshops for bone, stone, and wood had been established. Large crushing and chopping stone tools were discovered to have been used in the butcheries 400-300 tya. The same culture and workshop practices were performed at Neander, Germany 125-30 tya.

Acheulean Stone Age tools were discovered in a rock shelter in Jebel Faya, United Arib Emerants 125 tya. No human fossils were discovered.

In Germany art figurines of mammoth, horse, cats, bear, and bison were discovered with marked etchings scribed on them 32 tya.

In Florisbad, South Africa, a fossil skull was discovered with combined features of EEG, questionable EHEB, NEA, ADEG, and AS variant traits dating to 280 tya. It may be a specie variant far south from European population, because of variant conflict between the scientific experts arguing EHEB was begat from ARUD or EH in Africa. Others state ADEG and EEG were from Africa, and questionable EHEB, NEA, and AS were from Europe.

The features: primitive small brain skull: two times thicker than modern man with no crest ridge on the top, backward low sloped forehead, and protruding double arched bony brow ridges, face lacked a chin, large face muscles, and powerful jaws. It had adapted to the cold climate European environment.

H. Florisbad sapiens (FLBS) variants lived at Saldanha Bay Langebaan Lagoon, in Hedijiespunt 75 miles west of Langebaan Lagoon Cape Town, South Africa, with traits of ES, ADE, and early Modern sapiens variants EMS 260-150-125-100-40 tya. FLBS occupied Klasies River system caves to the exit at the Indian Ocean.

FLBS features were robust stocky and had a powerfully built body: skull; receding brow ridge approaching the features of EMS. Other fossils looked more like ADE and were similar to EMS from the neck down. Males were 5f 9i, 137-200 p, females 5f 2i, 112p.

FLBS were stone and bone tool crafters. They used sophisticated tools fabricated from bone, symbolic etchings, and stone point barbed harpoons, for their new hunting skills of fishing. The stone spear points were refurbished by sharpening the edges at Blombos Cave, South Africa 75-70 tya.

H. Klasies sapiens (KLA) EMS variant were beget from **FLBS** variant. Fossil records have included them as same species as EMS. They were capable of EMS thinking and behavior. They lived as cave dwellers 125-60 tya. They looked like EMS by 60 tya. They lived 350 miles east and north of Cape Town South Africa, at the Klasies River Mouth exit, at the Indian Ocean. Food was plentiful, but they were scavengers. They did not fish until 60 tya. They would not risk encounters with dangerous animals, but, adapted as advanced hunters and gathers, tool makers, fire makers at will in hearths, and symbolically used red ocher color in art. They would trade with each other.

THE FAMILY OF MODERN HOMO SAPIENS

The mutation of the brain of Modern Man over time resulted in strategic thinking and reasoning to plan and innovate. Modern Homo sapiens evolved as true <u>man</u>. They developed into the descendants from one single female living roughly 150 tya, in Africa. Homo genus mitochondria mtDNA of X linked through an unbroken chain of females were joined with a Y chromosome male. These are the specialized evolutionary changes in accordance with natural selection. Some European modern man do not have African ancestry, but Nordic and Scandinavian ancestry.

Civilization; a social culture composed by individuals who lived togagher, learning how to behave, and training their minds with knowledge, skills, and abilities to pass on to others who are civilized as a refined courteous culture.

Culture; groups of individuals who lived togagher, learn from their environment, language, technology, customs, and beliefs transmiting that knowledge, skills, and abilities from that generation to another.

Language: Genetic evidence passed on a gene associated with speech FOXP2 active in the brain and the specialization of the synchronous movements of the tongue and mouth necessary for speech. The mind recalls from memory (previously long term storage in various areas of the brain) many neuro electrical-chemical neurons through receptors in contact by synapsis in the brain processes, for body task action of the right combination structured or syntactical of the order of words, helping to determine the meaning of utterances that can be changed by the sounds and tone in the right order to communicate speech and ideas.

With the advent of 200 tya, Africa's climate started to become arid and colder. AS ADEG/LE living in Europe started their retreat to lower latitudes. Those who stayed adapted their bodies to the cold climate features, if they were not already adapted. Others migrated back to North Africa, settling along Egypt's Nile River System, on the south shores of the Mediterranean Sea to the Atlantic Ocean in Morocco, into lower Africa, and South Africa.

Early H. sapiens (ES) variant transitioned from AS. In Europe, Middle and East Asia H. sapiens form were manape 800 tya. They were known as, **"the wise man"**. They lived at Saldanha Bay Langebaan Lagoon; Hoedjies Punt, Langebaan, and Geelbek South Africa 250-164 tya. Other fossils were discovered along the Indian Ocean coast and inland caves in South Africa as late as 70 tya.

The fossils discovered at Herto, Ethiopia, 2 adults and a 6 year old child had primitive features a elongated skull with a large face. ES was transitioning toward modern humans and not fully modern until about 210 tya.

Early Modern H. Cro-Magnon sapiens (MCMS) were evolving in France and Spain. MCMS as apeman and were transitioning with features of early sapiens ES. They were man ape evolving at the Abri de Cro-Magnon overhanging limestone cliff rockshelter Les Eyzies de Tayac, in the Dordogne Region, in southern France 200-125 tya.

MCMS features: skull; brain 1200-1700 ml, high vertical forhead, small faced, forward protruded chin at the front of the lower jaw, rectangular eye sockets, small nasal opening, slender build, and tall. MCMS migrated into Spain 125-32-28-25 tya.

Europe was a cold climate 120 tya. MCMS retreated from the ice to warmer climate in southern Europe and southern Eurasia where the plants grew that fed the reindeer, ibex, and bison, which fed the humans.

Magdalenian H. sapiens (MMCMS) transitioned from MCMS/ES variant. Fossil evidence was discovered in southeastern France near the Ardeche River in a cave. MMCMS imagined animal spirits lived in the rocks at Salon Noir, France and used flickering torches as they entered caves 50 tya.

They had the traits of FLB KLA ES variants from Hegiiespunt Saldanha Bay/Langebaan Lagoon and the Klasis River system in South Africa 60 tya. They occupied caves in the Dorudogne River area 17-14 tya. Discovery of cave wall art paintings; bulls, horses, and deer, fabrications of; beads, fishing barbs, and stone points all fashioned around 24 tya. In other sites; in a cave in southwestern Lascaux, France and Altamira, Spain, paintings and art carvings were discovered. They were in search of the meaning of art. They developed paintings and sculptured man-beast figures in the same caves previously occupied by NEA 28-12 thousand years before the MMCMS occupation.

Optically Stimulated Luminescence determines when sediments contained in buried artifacts last saw sunlight.

MMCMS fossils were discovered at Beache Saint Vaast in northern France. Their hunting skills advanced to projectile weapons with finer blades. They survived on the animal population of bison, horses, red deer, and reindeer. They were dressed in animal hides 50 tya. Their brains adapted and developed to modern technology, Modern sapiens advanced thinking and dexterity by 38 tya.

Late MCMS variants spread into Europe from France were all man 50-40 tya. The NEA was challenged by the advance of the Late MCMS.

Early Modern H. sapiens (EMS) variant transitioned from ES, were all man evolving 164-154-80 tya. EMS had a combination of many features, traits, limited imagination, and no longer had a robust physique. Skull; changes in cranial bone structure with a protruding **frontal brain** averaging 1300-1400 cc, jutting jaw, high forehead, and a barely visible brow ridge discovered at Omo, Ethiopia, Africa dating to 125 tya.

Their neurological mutation markers led to transitioning to a spoken language and cultural behavior. There evolved a major change in the spine last vertebra of the thorax. The hole in the spinal bones center became larger to accommodate spinal cord nerves that control breathing and rib cage muscles used in exhaling the lungs. Males were 5f 9i 143p, females 5f 3i 119p.

Three EMS fossils were discovered dating to 164-154 tya. A child and two adult skulls were de-fleshed after death.

Living in northern Africa EMS migrated into the southeastern edge of the H. Neanderthalensis (NEA) territory along the Mediterranean coast to Israel 125-95 tya. The NEA encountered EMS tall humans living in their vacated caves in Israel 100 tya. EMS migrated into Europe, Eurasia, and the Middle East. The NEA reoccupied the vacated caves. There was no evidence the two humans interacted.

A skull was discovered in Luijiang, China of EMS dated to 100 tya. Fossil bones were discovered in Tiayuan Cave and Larbin, China dated to 40 tya.

EMS migrated out of Africa into east Asia 80-50 tya.

EMS/LNEA variants were discovered at the Portuguese site of Lagar Velho and at a Muierii site of Pestera cu Oase in the southwest Carathian Mountian of Romania dating to 32 tya. However, there were too few LNEA females on the landscape for reproduction.

Aurignacian H. sapiens (AMMCMS) transitioned from MCMS/ES variants evolving in France. It was the time of social cultural development, technology and art 28-25 tya. In southern France at what the locals called the, "Mas-d' Azil" gathering spot was where sapiens exchange goods, gifts, and found mates. There was another gathering spot called Niaux in the French Pyrenees.

Late Modern sapiens (LMS) transitioned from EMS and were all man evolving 95-80 tya. LMS brain was changing mentally and anatomically with a major shift in the frontal lobe; allowing them to be creative, with systematic thinking, and reasoning to plan and innovate. They had cognitive modern thinking abilities in **cultural buffering**.

Buffer is something in the group behavior; forming social origination or a cultural tradition or technology; skills preparing stone core, knapping long blade flakes, hedging its risks on high invested gains.

LMS emerged from southwest Asia, Europe, and northern Africa with a buffer of an economically efficient approach to hunting and gathering by team work resulting in a more diverse diet and the spread of the risk within the social unit.

The men were responsible to bring down large animals for meat. The women and children were responsible for small game and plant food. They needed only 2200 calories daily and enjoyed the benefits of their division of labor and living longer.

LMS social cultural sites represented larger populations as a unit having biological and social repercussions which demands more social interaction. There was a shift in burial style with the addition of objects with the body to assist in an after life belief. There was greater brain activity during childhood and adolescence, creating pressure for more sophistication of language. Longevity increased the life span of the group members. Intergenerational transmission of knowledge created cultural innovation of survival skills and tool making technology passed on to others from one generation to another and from other groups to others.

Time passed, many migrated north out of Africa into Europe and Asia and became cold climate adapted. The climate was cool and dryer caused by North Atlantic iceburgs cooling the ocean 90 tya. Some LMS migrated from northern Eurasia to southern Eurasia.

Fourteen modern looking fossils of NEA/LMS variants were discovered at Qafzeh near Nazareth, Israel 92 tya.

Many fossils of Modern sapiens from other groups from the Middle East region had LMS features 90 tya.

European Asian sapiens (EAS) transitioned from LMS. Central and south Europeans in the Middle East were LMS and Eurasian.

The EAS migrated to Mongolia, southern Central and Eastern Russia, and northern China. Part of the EAS group may have migrated south over Pacific Ocean islands exposed land bridges when the sea level was below

400 feet, to land of the Guinea 70 tya, and Australia 60 tya. EAS fossils were discovered in the Guenea's and Australia.

EAS separated into two groups. One group migrated into Central Asia skirting the mountains and desert reaching the Altai Region in southern Beringia (Siberia in Eastern Russia) 40 tya.

The other group of EAS inhabited Mongolia, crossed southeastern Eastern Russia and northern China. They migrated north over frozen Lake Baikal Valley, to the frozen Lena River Valley, to the Artic Circle at the Verchojansku Mountain Range in northern Siberia, and became isolated 32 tya. There is no verification these EAS migrated over Beringia to the Alaska Northwest Territory, Canada, and the Americas'.

EAS inhabited Serpentine Hot Springs in Northern Beringia, Siberia 12.4-12 tya. Some may have migrated east over Beringia to North America, others may have ceased to exist. Clearification has not been confirmed.

The 2nd group of EAS migrated east along northern China to the Pacific Ocean and became isolated by the Glacier Ice at the Pacific Ocean for 10,000 years. EAS mingled with the **Chinese H. pithecanthropus (EASPIS)** variant. EAS transformed as **EASPIS East Asians (EEAS)** variants by 30-20 tya. When the Glacier Ice started to retreat 21 tya it opened the Pacific Ocean passage north along the glacier ice over land bridges and frozen ice when the sea level was below 400 feet into the Alaska Northwest Territory in North America and the Cordillean Glacier. EEAS migrated along the Pacific Coast south into Washington State USA, on the west side of the Rocky Mountain Range. In the Americas' they were known as **Western Paleo American** variants in the USA. Their fossils were discovered on the islands of the Pacific Ocean at Los Angeles, California in stratum dated to 15 tya.

EAS found a way east over Beringia, Russia to the central southern coast of the Alaska Northwest Territory of North America about 19 tya. EAS lived in groups separate from each other in Alaska and transformed to the **First Native American sapiens (FNAS)** variant about 18 tya. Fossils were discovered in Alaska transformed to the FNAS variants.

Some of the FNAS variants continued risk taking migrating east into the Northwest Territory to the eastern side of the Canadian Rocky Mountain Range at Alberta, Canada. The east flank of the Rockies exposed a passage way south between the Cordillean and Laurentide Glaciers into valleys and land in northern Montana, USA. They migrated along the east side of the ice covered Rocky Mountain Range and Sierra Nevada Range to New Mexico about 16-12 tya. These FNAS variants were known as **Eastern Paleo American** variants in the USA. FNAS transitioned to **Native American sapiens (NAS)** variant about 10 tya.

EAS/FNAS/NAS and EAS/EEAS innovative creative values developed economic social technological diversity with the knowledge, skills, and abilities to perform tasks.

H. Anatomic Modern sapiens (AMS) transitioned from LMS and were all man living 80 tya. Their population was more numerous. AMS masses migration out of Africa into Europe 80-60-44-40 tya. There was a need for more sophisticated social and tool technology skills required to survive in Europe 75 tya. They were going through a mental transition of the brain frontal lobes. Their brain neurological system was better developed mentally, physically, cognitive traits, and technologically capable of Mind-Brain behavioral thinking in a modern way. Strategic thinking with the ability to abstract, analyze the past, anticipate the future, by reasoning; allows them to plan, create art, use a fully articulate speech to pass on information more efficiently building social culture, and develope new tool technology 70-60-40 tya.

The San People of southern Africa, Beaka Pymies of Central Africa, and some Ethiopian tribes spoke a language of tone sounds and clicks 70 tya.

Neanderthal NEA and AMS lived in a climate that was extreme and rapidly fluctuating. The climate was becoming colder 60 tya and many retreated south. AMS became more flexible, adaptable, and learned from experiences how to survive 50 tya.

AMS were hunters who used their balanced weighted throwing spear to kill larger animals. The spear stone spearhead tip was 7 to 10 inches long and was wedged into a green wooden shaft, lashed and locked in place with woven material, and dipped in hot Birch lacquer. The stone weight on the shaft behind the spear point was optional. These spears were made for killing and were more efficient with less risk of wounds to human.

AMS for the most did not allow social mingling and did not tolerate the lesser mentally challenged NEA. Each other were probably relatively swift and hostile excluding each other. Those determined by AMS with bad behavior or could not adapt to their modern social culture were not tolerated. NEA lived for the most as populations in exile in the shrinking pockets of habitation.

AMS applied simple concepts. This start of civilization and team work, sharing in social culture developed their technology; bone tools, bone needles to sew animal hides into weather resistant clothing they wore, crafted woven blankets and rugs from textiles, fabricated fishing and animal nets 300 foot long 3 feet high, fashioned figurines using stone tools, and used bone musical instruments 17-14 tya. They had a creative sense of prospective for artwork; expressing themselves in pictures, painting, and carving objects in search of meaning in their modern social cultural art. They discovered high heat caused the sand (silica) to turn into glass below their hearths. They crafted ceramic artwork and coatings. There were elaborate funeral practices performed for tribal social hierarchies. The penal system judged those with bad behavior or could not adapt to modern social culture.

On Sumatra Island, Indonesia the Toba Volcano eruption creating a 63 mile wide crater forming Lake Toba. The east winds deposited 6 inches of ash over southern India. Tools were discovered above and below the ash deposit at Jwalapuram, India dated to 70 tya ca. No fossils were discovered.

Off the tip of south India is the island of Srilanka. At Batadomba-lena tools were discovered dated to 40 tya ca, but no human fossils.

Under a protruding 30 foot cliffs AMS fossilized footprints were discovered in archaic dunes of gray sandstone lining the shoreline 75 miles northwest of Cape Town, South Africa, at Saldanha Bay Langebaan Lagoon dated to 60 tya ca. They were known as the, "**dune walkers**" foot prints. There were clear imprints of 10 ½ inches long foot prints resembling a Homo sapiens foot.

The foot balanced on one leg; heeiball firm strike, plyable flexable arch lift, alined the big toe and four assisting toes used for forward push off allowed the foot to propel the human forward in a stride.

Acheulean Stone Age Upper Paleolithic tools were discovered at a AMS Israel site and were identified as early technology used by the culture 40-30 tya. Tools were discovered and a fossil of AMS in Niah Cave, Borneo dated to 40 tya.

In the Czeck Republic near the village of Dolnii Vestonice and Pavlor,

AMS was cooking and heating with hot rocks, and fabricated ceramics of the Neolithic Period of time 28-24 tya.

Another group of AMS migrated from Africa into Europe 22 tya. Their population spread through Europe and Asia. They were not cold climate adapted. They lived in wooden tepee like structures lined with animal hides. Some had hearths inside their structure.

The Germanic and Slavic tribes in the far north were in a hostile climate and were overpopulated 20 tya. These tribes migrated south and southeast in Europe and the Middle East Asia. The migrations from Central Asia retreated west across European Russia into Europe. These tribes and families comprised of the rear guard of the migration and became the Slavic people in Europe.

In the village of Mezhirich, Ukraine, AMS shelters were built of mammoth tusks and bone in round foundations. Pits were dug in the permafrost to freeze the meat from the hunts. They may have been encouraged to settle giving up mobility and living by the rules of the tribe 15 tya.

H. Croatian AMS (CAMS) variants were **Eurasians** genetically identical to AMS 99.5% DNA. They migrated north into Asia to Czech-Slavic sites living in civilized communities 20 tya.

H. Moravian (MAMS) variants were **Eurasians** living in northern Asia at Czech-Slavic sites in civilized communities. They were cooking, heating with hot rocks, creating ceramic art objects; clay pottery, animal and human figurines, weaving; textiles, bags, blankets, clothing, and hunting nets fabricated to capture small game. Funereal ritual burial practices were performed. All these tasks were accomplished by cooperative team effort 20 tya. They belived and had faith in the sun and moon activity.

In the Nile River Valley of Sudan, at a site called Wadi Kubbaniya, conflict arose among AMS and the inhabitants 22-20 tya. There was confrontations and competition within the specie to survive. Fighting and killing broke out over the scarcity of resources. Massive graveyards were discovered dating to 14 tya ca. The killings were by spears and clubs. It looked like systematic warfare and cannibalization to survive. Woman and children were the venerable ones.

Violence was not in the Homo sapiens evolutionary legacy or in their genes. The learned aggression and defensive attitude escallated in a struggle for power over others. Once established as the dominate specie throughout the region, MAMS systematically held public executions as a control over non-conforming; vial misfits, ill behaved, and those who could not adapt or change were isolated, confined, or put to death 10-8 tya.

H. Indigent Aboriginal Australian sapiens (IAAS)/EMS/LMS/ AMS variant in southeast Australia along the Pacific Ocean at Lake Mungo, part of the Willanra Lake System, a fossil of IAAS/AMS was discovered dating to 60 tya. In the Northwest Territory tools were discovered in rock shelters at Malakunanja and Nauwalabia.

IAAS lived in caves along the Indian Ocean. They lived in a cave at Jinmium 58 tya. Stone tools were discovered dated to 176 tya ca. The cave walls were painted with ocher a high grade hematite crushed into powder. In a cave at Malakunanja II dated to 52 tya, ocher powder was discovered dated to 116 tya ca. They used the powder to color their bodies, in ceremonies, and designs. However, there is no evidence as to when the tools were fabricated.

Eurasian Theory: Eurasian/LMS (EAS) variant groups migrated from Indonesia south island-by-island over land bridges when seas were below 400 feet into New Guinea and Papua New Guinea, crossing at the Barrier Riff to Cape York Peninsula, Australia 60 tya. No evidence or facts support this theory. LMS live in New Guinea before 60 tya.

EAS LMS fossils were discovered along the Weber Basin shoreline in Northwest Territory at Cape Crocker, Australia in stratum dated to 60-50 tya ca.

Indonesian Theory: The migration crossing of Weber Bay over a series of land bridges from Jerimalai, Timor Island, Indonesia has no evidence or facts to support this theory. How EAS EMS migrated is not clear.

H. robustus AMS transitioned from IAAS variant evolving at Cohuna, Australia, 15 tya.

H. gracile AMS transitioned from IAAS variant evolving at Keilor, Australia, 13 tya.

Africa, Europe, Eurasia, Asia, and the Americas' begat: AH, EH, LH; AEG, EEG, ADEG, (AHEB?), EANT, LANT, NEA, ENEA, LNEA, CNEA, DCNEA; AE, EE, LE, ADE, AS, ES, EMS, LMS, AMS, MAMS; ES/MCMS, MMCMS, MACMS, IAAS/LMS/AMS, LMS Eurasian EAS, EASFNAS, LNAS, MNAS; EAS Eastasian PI (EEAS), and others created the diverse species of variants we share their genes with.

PRE H. NEANDERTHALENSIS
CONTROVERSY

The divergence, transformation, and transitioning from primate ape variants to man Homo genus species variant evolution are far from being clearly defined. Continuous research by scientists established the descriptions of diversity at different times within complex evolutionary events. Researchers are obcessed with their discoveries. They hasten to publish the new finding before they have true facts or enough fossil representation to support their subjective analysis concerning new specie as a convenience for discription and classification to their decleration claim to fame. Over time variations differ from place to place.

Archaic H. heidelburgensis (AHEB questionable existance) claim of its origin in Africa 1.9 mya. Controversy surrounds the origin of AHEB from Africa. Claim is AHEB variant was begat from early H. ergaster EEG, early H. erectus EE, and H. rhodesiensis. EEG migrated from North Africa into Eurasia and northern European Russia 1.1-1 mya.

There are claims by some researchers suggesting AHEB originated in Africa 1 mya-900 tya. They believe HEB beget from H. rhodesiensis (RHO) variant ancestor with traits to early H. ergaster (EG) and early H. erectus (EE) 800 tya.

NOTE: RHO variant with ADEG and EEG features are the subject of a controversy concerning the existence of A. heidelbergensis being RHO, but, has not been proven. There is no proff AHEB ever existed. No AHEB influencing relationship could have existed with RHO due to differences in discovery age and dating.

RHO features: skulls; discovered in the Kabwe cave system in northern Rhodesia, now Zambia, Africa were thick boned, large brain 1300cc, forward sloping skull cap, large arched brow ridge, horizontal ridge at the back of the skull. Researcher analysis attempted to prove HEB originated in Africa was arbitrated with their subjective determination. The researchers renamed the RHO skulls as HEB from Africa. This was a convenient discription and classification to their decleration claim to fame. The missing factual evidence open ends AHEB existance. RHO fossils had early H. ergaster EEG/early H. erectus EE variant ancestry. They were younger than the alledged age of AHEB. There is no evidence AHEB variant migrated north into Europe from Africa.

Early Antecessor (EANT) variant ANT means "ancestor", "Pioneer Man", had traits similar to NEA and had transitioned from EEG variant. EANT lived in the limestone hill caves of Sierra de Atapuerca in northwest Spain, nine miles from Gran Dolina and 31 miles from Burgos. These caves extend to more than 2.5 miles of limestone stratum dated to 1.1 mya-900 tya. The stratum from the Tinchera Del Ferrocarril trench is dated to 900-800-780-500 tya. Railroad workmen cut a trench through the southwest hills in the 1800's exposed sediment filled fissures. Olduvai stone tools hammer stones, small flacks, scrapers, were used to deflesh and woodworking. Many of the hominid fossils butchered and cannibalized displayed cut marks on their bones. The upper and lower jaw and teeth were complete, dated to 780 tya. The lower jaw was chinless. EANT was discovered by the team of Jose Maria Bermudez de Casto Spanish excavators. Their finds were of juveniles, adolescents, and females of about ten fragmented individuals. No complete skeleton was ever discovered. This was a convenience for discription and classification of Spanish declaration claim to fame.

The features of EANT were partical reconstructed composits from many individuals. It had broad long slender sholder callarbones and was wide chested. The skull; light bone rounded, brain 1000 cc or 61 cubic inches, fully developed modern middle face, high set cheeks, wide spaced eye sockets, teeth larger than modern humans and smaller in the upper jaw, lower jaw had a receding chin, arched brow ridges, prominate nose with a narrow upper opening. The neck vertebrae and ribs were similar to modern humans. It had a light bone structured, long limbs; forearms, wrist and hands, long thigh bones similar to NEA, and modern human feet. Based on the forearm bone they were 5f 8i tall. It walked with a modern gate. Age analysis was determined through biochronology and paleomagnetism to be 1.2 mya-500 tya.

Later Antecessor (LANT) variant transitioned from EANT and lived 580-370 tya. Other LANT variant fossils discovered in Europe had traits of NEA and were claimed to be living in Dusseldorf, Germany about 325 tya as NEA. There is no evidence that ANT migrated north from Spain. Here is another convenient discription and classification to LANT decleration lacking factual evidence.

H. heidelburgensis (HEB) variant claim was determined by a fossil of a large lower jaw mandible discovered in Grafenrain Sand Quarry Mauer, Germany, located on the Neckar and Rhine rivers junction by Otto Schoentensack. The jaw was sent to London, England for analysis and species designation. Researchers subjective analysis performed in London, England proclaimed a new specie name of **H. heidelbergensis (HEB)** dated to 600 tya.

Its features: had EEG traits, since there was no complete skeleton or skull evidence as researchers formed their opinions based on the thick bone, strong jaw may of had chewing muscles, and small molar teeth, with an estimated brain size of 1125 cc. There is no fact evidence for designating the partial skull jaw as new specie. This is a convenience for discription and classification to Otto's declaration claim to fame.

Claims of HEB discoveries in Asia; Ceprano in central Italy, and Petroloma Greece dated to 400-300 tya. The Spanish Scientists discoveries classified some of their discoveries of fossils were based on HEB designation of the jawbone partical skull determination. These dates are in conflict and may be 780/710/575 tya when some H. antecessor (ANT) variant transitioned toward what was classified as H. neanderthalensis (NEA). The composit skeleton of ANT may have HEB claimed EEG/EE bone fragments.

Determination of HEB used the jaw skull fragment as the reference basis for most of the future identification of discoveries of HEB in Europe. No factual evidence traces the ancestry of **Early HEB** alleged to be living 600-200 tya. Fossils were discovered above and below volcanic ash. Specie concept of skeletal remains of males claimed to be HEB were scattered in

Europe. In a limestone chalk quarry in southern Boxgrove, and W. Sussex, England of the United Kindom, two lower incisors of four teeth and part of a tibia shin bone of EHEB were dated to 524-500-478 tya. There was a discovered of Acheulean Stone Age tools; stone hand hammers, knapping of creative flakes accomplished various tasks of cutting flesh, preparing skins, cutting and crushing of bones for marrow, and woodworking.

In Arago Cave near Taulavel in southwest France, thick boned skulls and mandibles were discovered of claimed EHEB EEG/EE skull traits. Some EHEB migrated south of the Pyrenees into Spain due to the cold climate. HEB variants were discovered at Petralona, Greece, all dated to 400 tya.

In Sima de los Huesos (Pit of Bones) cave in northern Spain, at the base of a chamber, down a 45-65 foot vertical shaft, a discovery of 28 fragmented skulls are of mixed traits of EEG, ADE, and alleged EHEB, and a different early antecessor (EANT) variants all dating to 400-300 tya. These fossil bones were excavated and reassembled into what researchers classified as skeletons of the ANT variant species. Some researchers argued that ANT variant was early H. neanderthalensis (ENEA), with no factual evidence proving the claim.

The variants, for the most, exhibited features of thick bones ENEA: skulls; average brain size 1224cc, long forward low sloping skull cap, protruding large straight brow ridges, wide set eyes, flat face, large upper jaw projecting forward, large chinless thick lower jaw with bone extention for attachment of neck chewing muscles, small molar teeth, large nasal opening, cheek bones protruded for the attachment of jaw chewing muscles,

At the butchery in Belzingsloben, East Germany, claimes of EHEB or EEG variant broken bones of 37 fossils were discovered around an ancient calcium carbonate spring.

Near Hanover, Germany, at the Schoningen butchery, there was a discovery of spears that were not the hand held bayonet type. These spears were balanced throwing spears 6 feet 6 inches long. Near the town of Schoningen, Germany, in a peat bog strip mine a 7 foot long glistening dark brown throwing spear was discovered and was identified by the application of a balance weight to the spear, all dating to 400-300 tya.

No claims of HEB fossil discoveries have been disclosed living after 200 tya.

H. rhodesiensis (RHO) skulls were discovered in the Kabwe cave system of limestone metal sulphide ore mined in northern Rhodesia, now Zambia, Africa, and was evaluated as Archaic HEB by Authur Smith Woodward in London, England. No testing was performed. These skulls were dated using the relative dating method from other animal fossils of that period. This method is subjective and is unrealiable. This classification did not establish their claim to fame for archaic HEB African ancestry. However, the two claimed RHO skulls (EEG/EE) were redesignated as AHEB.

RHO features: skull; thick boned, large brain 1300cc, forward sloping skull cap, large arched brow ridge, with a horizontal ridge at the back of the skull. No other body parts were discovered.

The controversity: RHO fossils were early ergaster EEG early erectus EE variant beget 800 tya and were younger than the age of redesignated AHEB claimed skulls. Stone tools discovered at the find dated to 300-125 tya in rock stratum. RHO may have migrated into European Russia 700 tya, however, lack the evience of proff. RHO alleged presence in Heildelberg, Germany was determined to be AHEB from Africa, a convenience for discription and classification to the discovery claim to fame. No evidence has been presented that RHO was discovered in Germany.

There was another claim of a lower jaw discovered later in Germany claiming to be AHEB dated to 600 tya. No proff or fact evidence has clearified the claim was AHEB. However, the Spanish used the lower jaw as a reference basis for HEB future identification in Europe.

THE FAMILY OF HOMO
NEANDERTHALENSIS

H. Neanderthalensis (NEA) transformed from ANT variant as a manape. It was commonly known as, "Neanderthal", meaning, Valley of the Neander, located in a valley along the Dussel River, at Feldhofer Grotto Cave near Dusseldorf, Germany.

NEA genome has a three billion letter DNA sequence. A form of gene trait DNA genome sequencing MCIR endowed its carrier with red hair and pale skin pigmentation. The Norse Norway culture had the same gene traits.

DNA extracted from NEA bones sequenced nuclear (chromosomal) DNA has revealed 1 to 4% was found in modern Europeans and Asians. It was based on fragments of 100 base pairs dated to 40 tya. Some of today's people who came from Europe and Asia can trace back to NEA ancestral mtDNA of X in the cells of females who had daughters of now living humans from Africa. Some Europeans escaped the NEA DNA influence.

NEA bodies demanded high caloric intake of 4500-5200 calories daily, especially at the higher latitudes. They could not put on the fat insulation needed to combat climate fluctuations in the peak ice core maximum Ice Age in Europe 30-18 tya. They had a hard time surviving the cold climate, even though they had adapted to the cold during the winter months. They lived a life of hardships. Many of them would parish due to starvation only to be cannibalized by others. They scavenged and pilfered from others to live, and it was not uncommon for cannibalization. Woman and children were at a disadvantage when they were forced to join in the risk of rough and dangerous survival activities of the hunting.

NEA social units were no bigger than an extended family. Eight year old female fertile daughters moved out of the family group to another group to prevent inbreeding. Females were the ones who grubbed the ground for root food. Many lived their life of stress and died before they were 30 years old. NEA aged four times faster than the Modern H. sapiens. They did not enjoy the benefits of marked division of labor, as did the modern humans in Europe.

In Sierra de Atapuerca in northern Spain early heidelbergensis EHEB claim or H. ergaster ERG/early H. erectus EE, H. antecessor (ANT), LANT variants were living 800 tya. Claim of NEA fossils were discovered at Spy, Belgium in 760 tya stratum may have been misidentified. Fossils of EANT were found in Europe dated to 710-575 tya.

The pre-Neanderthal Early EANT variants were apeman and transitioned from claimed HEB or EEG/EE variant to manape. Late Antecessor (LANT) variant was transforming to early ENEA 580-370 tya as a manape.

At the Bitzingsleben butchery site in eastern Germany a verity of artifacts and 200,000 stone tools of hand axes, and hand held three sided cutting-scraping-crushing hammer stones were used until 250 tya. 100 large and small animal bone, wood, antlers, and 37 LHEB questional claim or EEG ADE/ANT/NEA variant species fossils were preserved in calcium carbonate dated to 400-380 tya. Discovery of ENEA fossils from Neander and Dusseldorf, Germany, in stratum were dated to about 370-50 tya ca. NEA transitioned as all man and lived as cousin to Homo sapiens man in Europe until the NEA were extinct 8 tya.

NEA fossils were discovered at Saccopastore, Italy 120 tya; in Tabun, Israel; Gibraltar Caves in Spain 100-26 tya; Teshik-Tash in N. Uzbekistan 60 tya; LA Chapelle-Aux Saints, France; Amud, Kebara, and Shanidar, Israel 50 tya; Saint Cesaire, France 37 tya and Chatelperron, France 33 tya; Mezmaiskaya on the Black Sea 29 tya; Krapina Croatia 28 tya; were extinct in western Spain 28 tya, Lagar Velho Spain and Zafarraya Cave 27 tya.

In the Sierra de Atapuercas Mountain Range in northern Spain, near Burgos, located in the back of the cave of La Sima de los Huesos (the pit of bones), entered from ground level, down a 45-60 foot vertical shaft to the base, the interior slopes down about 100 feet in about a half a mile, and is where other human fossil fragments were discovered. Excavation of the contents of this pit; twenty eight skulls exhibiting EEG, ADE, EHEB claim, and EANT variant traits were dated to 400-300 tya. ENEA fossils from the secondary cave pit dated to 350-300 tya.

NEA living in northern Europe were sparsely populated 200-170 tya. The climate was another severe cold period 189 tya. In the north the fluctuations of warm-cold extremes caused some to migrate south. NEA numbers were reduced to below a sustainable level and were extinct in southern Spain around 28 tya.

Buffer; is something in group behavior forming social origination or a cultural tradition or technology; experience, knowledge, skill, and ability. Examples; Preparing stone core, knapping long blade flakes, and hedging risks on high invested gains in large animal hunts.

NEA society differed from Homo sapiens in a way crucial to group survival by cultural buffering.

Early NEA sapiens (ENEA), were transitioning from apeman to manape. Their population was overlapping other sapiens in Europe and Asia. They lived throughout Europe, Middle East, and Asia where few fossils were discovered 350-120 tya; Levallois, France 300-190 tya; Atapuerca, Spain 300 tya; Steinheim and Mauer, Germany near Heidelberg on the Rhine River 250 tya, Ehringsdorf, Germany 230 tya; northern Europe British Isles and Pontnewyold, Wales 250-225 tya; Beache Saint Vaast, France 196-159 tya; Krapina, Croatia 130 tya; Saccopastone, Italy 120 tya; in the Middle East at Teshik-Tash Uzbekistan; Amud Israel; southeastern Mediterranean coast and in Asia; Republic of Georgia (south Central Russia); northern China; and the only African NEA fossils were discovered at Neckar in central Africa. NEA started to transition form apeman, to manape, and were cold weather adapted in body form 250 tya. The declining NEA population was due to severe cold and the increase of early Modern sapiens (EMS) 120 tya into Europe.

About 180 tya, ENEA slowly adapted to tools. At Vezere Valley in southwestern France, hand axes were discovered with three functions notched in the stone that fit the hand for various tasks. Scares on an ax blade indicate they made sharpening repairs to the cutting edge. Hand axes were multi-purpose; cutting flesh, scraping, and crushing bones to extrude the marrow to eat. The handle version of the ax was used as clubs in high risk hunting of big animals. They advanced to hand held bayonet type wooden sharpened spears for thrusting into the animal. There were many hunting injuries with a high count to the arms and head. They had not developed projection spear skill for the throwing spear to kill large animals, which would have reduced injuries. Lacking the hunting tools they relied on cooperation, surrounding and confusing the prey for the kill, which is a predator tactic. The women and children would drive the prey from behind into an ambush from the men ahead of the prey for the kill. Work task division had taken place.

ENEA lived in the **Croatian** village of Krapina 130 tya. Their features: skull; sloped back low over their spine, large brain without the frontal lobe development for critical thinking, high forehead of modern man, broad nose, face juts forward lacking a strong chin, cheek bones were angled to the side and not forward. They lived in caves or rock cliff shelters and were clothed in animal hides not sewn together. They made fire at will, and cooked meat. They rarely made tools crafted from bone or stone.

ENEA were brutal hunters, resourceful carnivores, surviving in a cold environment. ENEA was unable to plan or organize. They were living in territorial hunting groups, sparsely and widely disbursed in small tribes, nomadic, and followed the animals in small groups of about 20. There were territorial disputes, which may have met with mortal combat. The old people were excluded. Keeping up with the nomadic move was a must for survival. They would setup temporary encampment with shelters made of wood and skin.

ENEA were primarily meat and bone marrow eaters. The bones found in Neander Valley, Germany, have cut marks from stone tools found in large butchery areas where dismembering of humans and the cannibalism of their own was conducted. They used fire; cooked the meat to destroy parasites and bacteria, provided warmth, and to act as a deterrent to predators and scavengers. They did not have hearths lined with rocks, but, campfires one built on each subsequent fire, ash on ash.

ENEA could speak, but did not develop a complex language, or symbolic and cultural consciousness. They carried a version of the gene FOXP2 associated with language ability active in their brain, and some of the vocalization. The range of vocalization may have been restricted. The hyoid bone located in the throat is associated with speech. It articulates speech sounds and may have produced a high pitch compared to modern man. They did not have structured language capable of communicating ideas and thoughts. They used signals for basic needs similar to infants when they start to speak.

In Hungary there was a discovery of a **hyoid bone** or voice box. This discovery indicates ENEA may have spoken a rudimentary language within a primitive social organization 50 tya.

ENEA features: had strength and endurance, short stature, powerful musculature body, thick bones, amount of skin area compared to their body volume was needed to conserve body heat (**cold climate adapted**), forearms and lower legs were short compared to the upper arms and legs, large hands, muscular fingers with a powerful grip, skull; had a protruding brow, was rounded with a back sloping forehead, and a low brain cage 1100-1400 cc, wear on their teeth indicated they often used their teeth as a vice to hold one end of a skin while scraping off the fatty tissue and then chewed the skin to make it supple, the nicks on the teeth indicated most of them were right handed. They did not develop sewing of hides with bone needles, absents of bone needles suggested they did not sew their clothing. They may have used stone tools to perforate materials. Males were 5f 5i 185p, females 5f 1i 176p.

In southwestern France, at La Ferrassie Mousterian site the ENEA occupied caves 180-60 tya. They were still using primitive tools 40 tya. Some of the ENEA started to advance their tool making skills, patterned after Modern H. sapiens.

There was another decline in NEA population before 125 tya.

About 120-100 tya the northern early and late Modern sapiens EMS and LMS variants from Africa migrated and encroached on the southeastern edge of the LNEA territory along the northern Mediterranean coast to Israel. The LNEA encountered tall EMS and LMS living in their abandon caves at Skhul and Qafzeh, Israel. When EMS/LMS migrated into Europe the NEA reoccupied the caves. There was no evidence they interacted. In Europe Anatomic Modern sapiens (AMS) systematically excluded the LNEA from their social culture because of their primitive negative attitude to changes. In isolation the NEA had a hard time surviving.

Stone Age tools were discovered in small Skhul caves dating to 120-100 tya ca. At Mount Carmel Hills southeast of Haifa, Israel 10 fossils with late LEMS features and one female LNEA were discovered in Tabun Cave, dated to 110 tya. Their brain size averaged 1518 cc.

In a cave at Shamidar, Iraq on the NEA major migration route into Asia, three fossil skeletons were discovered dated to 100 tya. Before the event of the use of thrusting spears there was a high risk of being injured during the hunt of large animals.

Fourteen modern looking fossils of ENEA/EMS/LMS variants were discovered at Qafzeh near Nazareth, Israel and at another site of NEA fossils exhibited a bony chin and a high forehead, all dated to 92 tya.

There may have been some **social behavior** concepts starting to emerge. This means their **brain frontal lobe** had developed **emotion and compassion.** Care for three individuals, injured in the hunt, was provided by others. Examination of the fossils shows healing of wounds they could not have survived without care during a time when the group was not on the move. They were fed well, protected, and helped to move.

Late NEA sapiens (LNEA) transitioned from ENEA and were all man 140-100-80-40 tya. They populated Europe and Eurasia (European Russia) 90 tya, southwest Asia 100-40 tya, and Europe 50-40 tya. LNEA was going through behavial and cultural changes. They were Pagans and belived in the after life and burial. Art was accomplished in some caves for ritual and magical purpose 50 tya.

Some of the LNEA migrated to Belgium. The skull of a child was discovered in Scladina Cave, Belgium dated to 100 tya.

The La Quina LNEA skull was discovered in France 75-40 tya. Fossils from Kebara were dated to 60 tya. LNEA migrated to the Liberian Peninsula into Budapest 50 tya. A fossil was discovered 40 tya at Amud, Israel exhibiting a brain size of 1740 cc.

Denisove NEA fossil finger bone was discovered in Southern Siberia, Eastern Russia 200 miles north of a point where Mongolia, Kazakhstan, and China borders meet, in the Altai Mountains, in the Cave of Denisova. The discovery was located in a back chamber stone layer dating to 50-30 tya ca. It was determined by DNA analysis from 70% endogenous (growth from within) DNA without male chromosome, belonging to an 8 year old girl of NEA ancestry. No other genome sequencing from the girl has been found that influenced any population from the surrounding continents. She may be the last of her kind LNEA or CNEA. Also, two human large molar teeth were discovered in the same cave and were DNA analyzed at Novosibirak, Russia, to be human. The molars were not NEA and resemble the teeth of AFR1/BOIP/AETP variants from North Africa or may be LMS EAS.

The European Synchrotron Radiation Facility generates x-ray beams in a particle accelerator that penetrates through the subject being examined dated to 10 tya for **neonatal stress** lines of the life of living moments of an individuals environment captured in bone, like teeth. Nanometer markers leave distinct marks on developing bone (growth rings).

LNEA was challenged by the advance of LMS and AMS variants spreading into Europe from Africa 50-40 tya. LNEA did adapt to tool making by "knapping" core stones ahead of the striking action produces sharp edged tapered pointed tools. Stone tools were used in the animal butchering of bears, aurochs (giant cattle), and longhorn oxen cattle. They used fire to cook, heat, light, and for protection. Their brain had not adapted to creative thinking about art. They decorated their bodies with manganese dioxide black pigment.

The few NEA may have mingled with LMS 44-40 tya, but, for the most were disassociated socially. In Pitrolona, Greece thick boned skulls were discovered with features of EEG, ADE, HEB questionable claim, and CNEA variant traits were evolving in southern Europe.

NEA fossils were discovered in Spain at Gibraltar Gorham cave and at Forbes Quarry dated to 28 tya. These caves were occupied by ENEA for 60 thousand years before 140 tya, from 200 to 140 tya. The LNEA lived there 140 to 80 tya. AMS occupied these caves 80 tya. Some of the LNEA tools were dated to 29 tya ca.

LNEA beget with EMS discovered at a site called Muierii in Romania 32 tya, and in a Portuguese site at Lagar Velho 30 tya. However, there were too few reproductive females on the landscape to sustain successive generations.

Classic Neanderthal sapien (CNEA) transitioned from LNEA and were all man throughout Europe 125 tya.

CNEA features: adapted changes in the bodies discovered from the Mladec site in the Czech Republic were thick boned, robust ribcage (barrel chest), with considerable mass of muscle, and they were strong, skull; had a noticeable domed skull roof which protrudes out slightly with a bulge at the base of the skull, known as a "Occipital Bun", the lower jaw was internally strengthened at the front so there was no chin. They learned to fabricate wooden spears with stone flakes mounted to the tip end of the shaft, bound with green tree fiber bark, and sealed with hot pine tree resin heated over a fire 40 tya. There was a shift in burial style by adding art affects to support their spirit belief in a after world 30 tya. No spirit had ever disturbed the grave or its contents upon discovery.

CNEA/LMS variants were discovered in Liberia 40-25 tya. These fossils resembled the anatomy of a LNEA having traits of LMS with evolved cold adapted limbs. They were a mosaic form of uncommon mingling of CNEA/LMS.

The falling temperatures 35 tya caused a decline in the food source. With 87% increase on a single source of food, the reindeer, surviving was problematically at risk. It forced the remaining NEA to flee south into Spain at Costa del Sol, the cliff sea side caves called Zafarraya Andalusian, and in the country side in northern and southwestern France along the Atlantic Ocean. Fossiles of jaws and femurs were discovered to have lived about 33 tya.

At Goats Hole Cave on the Grower Peninsula coast in South Wales the sea level was low 28 tya. The climate deteriorated and it was very cold. The NEA population in northern Wales, Scotland, and the Pennines mountains were extinct by 23 tya, when the **Pennines Ice Age Sheets** moved from the European north peaks to the south 22-20 tya. At Gibraltar Gorham caves in Spain AMS arrived and occupied these caves 20-19.5 tya.

A combination of environmental unrest, increasing competition for food, and the mass advance of MMCS into Europe 60-40 tya caused a decline of NEA. There were few NEA living 30-28 tya. Masses of AMS migrated into Europe from Africa 22 tya. The NEA was extinct by 8 tya.

NEA **Uluzzian Culture** in Italy and Greece improved their tools after the advance of modern humans 20 tya.

FAMILY IN THE
AMERICAS'

This was a time when deglaciation of the ice sheet over Europe, Asia and North America started about 21 tya when sea levels were below 400 to 300 feet. How they migrated into North America needs clearification. Theories of migrations and groups of EAS and EEAPS had to have migrated over Beringia, (Siberia), in Eastern Russia, into North America; land now Alaska, Northwest Territory, Canada, and the land before there was the United States of America (USA). Some of the EAS variant groups did migrate from isolated areas in the Eastern Russian mountains southeast onto ice barian Beringia and central southern Alaska. Europeans and Middle East groups living in Asia were Eurasians Modern sapiens (EAS). They transitioned to the **First Native American sapiens (FNAS)** variants living in Alaska about 16 tya. A group of FNAS migrated east over the Cordilleran Glacier to the Rocky Mountains in Alberta Canada. They discovered a passageway south into land in now northern Montana, USA. Some did survive the venture proven by their fossil discoveries in Alaska, Northwest Territory, Canadian Alberta passageway, and in northern Montana USA. EAS/FNAS variants were **Eastern Paleo American** in a new land transitioned to early ENAS and later to modern MNAS in what became the USA, Mexico, Central America, and South American cultures.

EEAPS variant groups were transitioned EAS Euroasian/H. pithecanthropus sapiens variants from north China who may have migrated along the Pacific Ocean coast onto ice covered Beringia and Alaska along the Cordilleran Glacier Pacific coast into Washington State, USA, on the west side of the Rocky Mountain Range. These migrations lack complete facts and need more clarification. Few human fossils and tools were discovered along those routes dating to 18-15 tya. EEAPS were **Western Paleo Americans** in the USA.

The Artic Northwest Migration Theory: the 1st group of EAS may have migrated south along the Artic Ocean coastal range south when sea levels were below 400-200 feet or may have skirted around the mountains into Beringia and the Alaska Northwest Territory (the Bering Strait ice land bridges into the Alaska Northwest Territory).

Part of the first group of EAS became isolated at the Verchojansku Mountain Range at the Artic Circle 32 tya. Another group of EAS became isolated at Serpentine Hot Springs in Northern Beringia 30 tya. These EAS may have survived the cold climate and may have migrated southeast through valleys into Beringia and Alaska Northwest Territory 15 tya. No fossil evidence or facts have confirmed this theory.

A group of EAS in Northern Beringia, Siberia East Russia habitation was discovered and may have migrated 12.4-12 tya over Beringia into Alaska needs better clarification.

Who and where these groups migrated from need better clarification.

EAS Beringia and Bering Strait Alaska Northwest Territory Canada Theory: EAS sites of habitation were discovered; some lived in southwest Alaska and transitioned as **EAS North American First Native sapiens (FNAS)** Tuluaz, Nogahahabara, Old Crow, Bluefish Caverns, and Swan Point, Alaska about 16 tya. The two North American western Cordilleran and eastern Laurentide Glaciers provided a interior passageway north to south in the Canadian Rocky Mountain Range (RMR) at Alberta, Canada 15-13.5 tya. A group of EAS/FNAS variants migrated east over the Cordillean Glacier to the north part of the RMR at Alberta, Canada, south between the Western Cordilleran and Eastern Laurentide Glaciers passageway, along the flanks of the glaciers and high mountains, into the valleys, and below the maximum ice sheet and land into Montana, USA 12 tya. There was evidence of human habitation at the center of the corridor at Charlie Lake Cave site, Alberta, Canada 12.35 tya. How EASFNAS migrated has not been clarified.

These EAS/FNAS transitioned to **Early Native American sapiens (ENAS)** variants in the USA without the EEAS influence 13-10 tya.

ENAS was discovered in Central America. In a submerged cavern

Hoyo Negro, "Black Hole", in Yucatan, Mexico a nearly complete skeleton of a teenage girl, "INAH or Nasa", named after the Greek water nymph, was discovered by sea divers and excavated from an underwater stratum abyss dated to 12.8 tya. This cave was a dry cave 10 tya.

The Pacific Coast Theory: There is evidence of EEAPS habition in North America below the glacier maximum ice sheet. These EEAPS variants lived along the Pacific Ocean and the west side of the Cordilleran Glacier covered by the RMR and Sierra Nevada Range (SNR) in North America, as Western Paleo American sapiens.

LMS/EAS European East Asian from the second group migrated east to Sichote-Alin, China 25 tya. They became isolated there for 10 thousand years to 15 tya. The **pithecanthropus (PI) sapiens** variants were also living in Sichote-Alin. The **European East Asian** EAS/PI became the EEAPS variants.

The glacier ice sheets opened along the Pacific Ocean when sea levels were below 300 to 200 feet. There were EEAPS risk takers 15 tya. EEAPS migrated from north China at the Pacific Ocean to Beringia into the Alaska-Canadian territory and west side of the Cordilleran Glacier into North America along the Pacific Coast 15-14.5 tya is questioned. EEAPS may have migrated from island and ice patches by following the Pacific Ocean shoreline following the sea mammals and fish along the Alaska-Canadian Cordilleran Ice Glacier to the maximum southern glaciation of open land in North America United States of America (USA), what is now Washington State, over routes along the Pacific Coast, lacking the needed facts and clarification.

The skull of a male EEAPS was discovered at Horn Shelter, Texas dated to 12 tya. The features: skull; large, longer, narrow, and less rounded, large eye openings, high cheek ridges extending to the back of the skull, protruding brow ridges, broad nose, and a protruded upper face. EEAS fossils were discovered at the Palsley Caves site and Columbia/Klamath River sites in Oregon, USA and the Channel Islands sites in California dated to 14.35 tya ca.

NATIVE AMERICAN SAPIENS

The Laurentide Glacier Ice covered Eastern North Americas 15-14 tya. EMS/FNAS migrated on the east side of the ice formations along the RMR and the Sierra Nevada Range (SNR) into northern United States to

New Mexico. EMS/FNAS were living in La Serra, Lovwell, and Clovis, New Mexico. They inhabited; Debra L. Franklin site at Austin, Texas 15.5-13.2 tya, with claims to have been the earliest inhabitants in the Western Hemisphere.

FNAS in the USA were the Paleo Americans at the Meadowcroft site in Ohio. There was fossil evidence at the skirt of the Laurentide Glacier Ice Sheet that covered eastern North America 15-14 tya. They migrated to the Atlantic Ocean via the Great Lakes.

Groups of EMS/FNAS migrated east along the ice maximum deglaciation in what in now USA northern border with Canada into the Schaefer/Hebior site in Illinois 14.8-14.5 tya, Linsay site in Idaho, and Meadowcroft site in Ohio 14.25 tya ca.

A group migrated to the Atlantic Ocean and south to Page-Ladson site in Florida 14.4 tya. Others migrated around the Gulf of Mexico into northern South America to the Atlantic Ocean.

The discovery of a human fossil at the Manis site in Washington State was dated to 13.8 tya. It has EAS features, not EEAPS features.

A full fossil was discovered at the Anzick site in northwestern Montana dated to 12.65 tya. There was remanence of glacier formations along the RMR and south extension of the SNR in a line from La Sierra, Lovwell, WA, eastern Oregon, to Clovis, New Mexico.

The Clovis Culture of people fabricated Chert stone spear and blade point 11 tya. They hunted the bison, mammoths and other animals. At Folsom, New Mexico a cowboy in 1903 found the remains of a Mastodon dated to 10 tya.

Close observation disclosed spear points among the bones. Spear points have been discovered in this area in stratum dated to 13 tya ca.

EAS FNAS variants transitioned as **Late Native Americans sapiens (LNAS)** 13-10 tya. Analysis of ancestral mtDNA of NAS is lacking due in part of too few fossil discoveries.

As time passed they became more settled with physical and behavior changes in groups of tribes over 500 languages in various locations as they transitioned to **Modern Native American sapiens.**

There were two kinds of NAS. The farmers lived in villages in tribal culture settlements and the hunter wanders migrating on constant drive after the herd animals, raids and war on other tribes, and war with the USA government over land and treaties.

NAS societies were whole democratic settlements, others had very ridged class systems based on property, some were ruled by spirit gods, some had judicial systems punishment by torture, some lived in caves, others in tepees of bison skins, and cabins.

Tribes were ruled by worriers or by women, elders or councils, fraternitie ritual and membership unknown to the rest of the tribe, and tribes that worshiped the bison or matriarch. There were tribes that did not know of war. Other tribes were at war for centuries of fighting.

The South American NAS inhabited the Pacific Coast at Monte Verde, Chile 14.5-14.25 tya. Others inhabited Middle America, northern and southern coast of South America to the Arroyo Seco 2 site Sumidouro Cave, Brazil where a fossil of "Luzia" was discovered in stratum dated to 14-13 tya ca.

FNAS NAS variant complete fossil was discovered from the Anzick site in western Montana dated to 12.65 tya. It was conclusively determined from new unique markers by DNA analysis that is only found in Modern NAS. There is no Asian PI ancestry, suggesting NAS are related to the European LMS EAS FNAS. Two infant fossils were discovered in northern MT. Their DNA was confirmed to be related to European FNAS. When the DNA genetic markers evolved has not been clarified.

The complete fossil of EASFNA variant known as the, "Kennewick Man", was discovered along the Columbia River at Richland, Washington state, USA dated to 9 tya. It was first identified as Caucasion with European traits, not Asian. It had chest wounds. DNA testing and analysis have been conducted by the Smithsonian Institute who argued the man was not NAS. However, it was confirmed to be most closely related to Modern NAS (MNAS). MNAS argued he is of their culture the, "Ancient One". The fossil was housed at the Burke Museum at the University of Washington. Legislation signed by President Barack Obama 12-19-2016, transfers the remains of bones to MNA Confederated Tribes of the Umatilla Indian Reservation of Colville, Bands of the Yakama Nation, Nez Perce Tribe, and the Wanapum Band of Priest Rapids. EASFNA was reburied at an undisclosed site.

Chart 3.
Phylogeny Of Hominid Apeman Manape Man 900-10 Tya

Primate Population Loc.**	Species Variant
CAN USA C/MA SA	Advanced Native American H. sapiens (ANAS)
USA	10 Modern Native American H. sapiens (MNAS)
AL CAN	13 Late Native American H. sapiens (NAS)
AUS	14 First Native American H. sapiens (FNAS)
USA	16-10 Paleo American (PA), Eastern EAS/FNAS, Western
EAS/PI (EEAPS)	
CZ CR CZR CS	20 Eurasian-Croatian H. sapiens (ECS)
NG CH RUS	20 Eurasian-East Asian pithecanropus H. sapiens
(EEAPS)	
CZR	20 Mod Moravian H. sapiens (MMAMS)
FR	28 Mod Auriracian H. sapiens (MAMCMS)
(EURO RUS) UZ AL USA	32 European Eurasian H. apiens (EAS)
FR SP	50 Mod magdelenian H. sapiens (MMMCMS)
AUS	60 Indi Abri Australian H. sapiens (IAAS)
AUS NG UKR IS CZR CS CZ SP SA KN SU MO 80 Anatomic Modern H. sapiens (AMS)	
AUS NG IND CH RUS FR AL	95 Late Modern H. sapiens (LMS)
IND	95 H. floresiensis (FLO)
UKR IND RUS SY IRA IS CR YU RO HU BE 100 Late NEA (LNEA) PO FR SP IN	
IND	100 Advanced pithecanthropus H. sapiens (ADPIS)
IND	100 H. soldonsis (SOL)
LI GE SP WA	125 Classies H. NEA (CNEA)
FR SP	125 Modern Cro-Magnon H. sapiens (MCMS)
IS CZR CZ RO EY RUS FR SP AUS SA UZ 164 Early Modern H. sapiens (EMS) KN ZI SU ZA AL MO	
FR SA	250 H. klasies (KLA)
YU IT GR BE FR SP UK WA SU	250 Early H. sapiens (ES)
IN CH TU AL	250 Late pithecanthropus H. sapiens (LPI)
FR SA KN	260 H. florisbad sapiens (FLB)
NK	260 H. jinniusham sapiens (FLB)
ZM/RHO	300 H. rhodesiensis (RHO) ADEG*
NK IND ET	300 Advanced H. erectus (ADE)
LI CH RUS IT GR SP UK SC 370 Early H. neanderthalensis (ENEA) WA	
SP	580 Late Antecessor (LANT)*
GR SP BI UK	600 Early H. heidelbergensis (EHEB)*
700 Early pithecanthropus H. sapiens (EPI)	
NK CH	750 Late H. erectus (LE)*
CH ET	800 Advanced H. egaster (ADEG)
SP ZM ZI	800 Early H. Antecessor (EANT)*
MO ZA AL AN 825 Archaic H. sapiens (AS) LI TU TA SU ET KN ZI SA WA UK SP FR BE GR IT	

Time tya 9 8 7 6 3 2 1 90 80 60 50 30 20 10
Hominidae

* See text for Pre H. Neanderthal Controversy.

** See Primate Population location of species chart 4; African, European, Asian, Australia, First Native American, East and West Paleo American, Native American, Modern Native American, Advanced Native American.

THE GREAT NOMAD
MIGRATION THRUSTS

Over millions of years the nomad gene trait to imigrate was part of these populations from Africa, Europe, Middle East, Asia, and the Americas'

The geographically divergent populations are African, Asian, Caucasian (mostly European), Australian, and New Guinean, from several genetic sites for each individual. Of the total mtDNA of 11%, ½ of the variations fell into two groups. The first group sites were only Africans. The second group sites were for all other individuals plus some within Africa. The bulk of genetic diversity was found in Africa, taking the longest time in human genetic female-female ancestral information. From various populations gives us an idea of where and when groups diverged their ways in the great migrations.

NOTE: The horizontal latitude of angular distance in degrees around the Earth is measured from the equator zero in each direction north and south to the poles. The vertical longitude of angular distance around the Earth is measured from pole-to-pole zero to 180 and back to zero degrees in either direction east or west.

The African continent contained solid matter from the east and north and was connected to land in southwest Asia and the Indian Peninsula 18 mya.

AFRICAN GROUP

The continent south of the Mediterranean Sea is located 15 to 50 degrees north longitude and 35 degrees north latitude to 2 degrees south latitude. Migrations occurred from Egypt, Libya, Tunisa, Algeria,

Morocco, Sudan, Chad, Ethiopia, Somalia, Tanzania, Kanya, Zaire, Uganda, Malawi, Mozambique, Zambia (Rhodisia), Zimbabwe, and South Africa. Out of Africa there were many migration major thrusts into Europe, Middle East, and Asia 1.9 mya-500 tya.

AE and EE groups did not migrate into Europe settling along the south Mediterranean shoreline of the Atlantic Ocean. They started their journey in several groups over time from Ethiopia following the Nile River System, north around the east end of the Mediterranean Sea. They hunted animals on the savanna and fished for large salmon from the Nile River. EH/LH, AEG/EEG/ADEG, AS, and others migrated north and east out of East Africa, from Egypt.

LH, AE/EE and AEG/EEG migrated out of North Africa eastward into Asia 1.7-1.6 mya. This population split and others followed the coast around the Arabian Peninsula into India and lower east China. Some of the population each generation migrated a few miles further. Others stalled temporarily in Eurasia Middle East and others migrated southeast into Java and Bali, Indonesia. The migration was stopped at the great deep Java Trench waterway barrier with its treacherous currents from the Indian Ocean prevented migration into Australia.

EEG groups migrated from Africa into Europe 1.1-1 mya. Near the Sea of Galilee, in Israel, stone hand axes were discovered in stratum 1.4 mya ca. Both groups took their tool making knowledge and experiences with them 1.7-1.37-1 mya.

LE last migration thrust out of Africa was into the Middle East and Asia. The migration thrust of 700 tya took their tools and knowledge with them, and is the reason why no axes or tools have been found in Africa after that date.

Groups of Homos migrated west to Cape Horn, Tunis, Tunisia, Africa 1.6 mya to 700 tya. EE/LE and AS migrated further west along the south shoreline of the Mediterranean Sea living in Morroco and the Alantic Ocean side 125 tya. They were still living in this area 80 tya. H. erectus did not migrate into Europe.

A group of ADEG settled in various areas in the Middle East and Asia and became cold climate adapted. They migrated west from China to the Middle East, Europe, and along the south shoreline of the Mediterranean Sea, in Africa 600 tya. Some of the population each generation migrated a few miles further. Others stalled temporarily in the Middle East, Central Asia, northern China, and southern Central and Eastern Russia. It was very cold in Asia when AS, EMS and others migrated north into the Middle East Asia, Republic of Georgia, and Uzbekistan 500 tya.

Some of the South Africans AMS migrated north along the Atlantic Ocean to the Mediterranean Sea and east to the Pillars of Hercules. **H. florisbad sapiens** (FLB) nomads from South Africa who migrated along east Africa joined AMS. Some from this group migrated west to Algeria on the Mediterranean Sea. The FLB ADE and AMS groups migrated north along the western coastal plain to the Iberian Saudi Arabian Peninsula and north into Europe toward Scandinavia, where they became cold climate adapted 150 tya.

EMS groups began their migration north out of Africa 120-100 tya. The EMS Israel group migrated to the eastern Mediterranean where they encountered NEA 90 tya. EMS was clothed with animal sewn hides. They became cold climate adapted and lived in wooden tepee like structures lined with animal hides. Some had hearths inside the structure.

ADMS migrated out of Africa 50-40 tya. ADMS systematically excluded the NEA from their social culture, because of their primitive negative attitude toward positive life change values.

EMS AMS were in northern China and along the southern border in Eastern Russia 40 tya.

Homo populations grew along the Nile River from those who had stalled in Asia and migrated back to Africa. In the Nile River Valley of Egypt, at a site called Wadi Kubbaniya, conflict arose among the ADMS and other inhabitants 22-20 tya. There were confrontations and competition between the species to survive. Violant fighting and killing were learned ill behaviors. Their emotions created stress over the scarcity of resources for survival.

In Sudan mass burials were discovered dating to 14 tya. The killings by each other were with spears and clubs. It was systematic killing and cannibalization to survive. Women and children were the venerable ones.

Some of the South Africans ADE, ADMS migrated north along the Atlantic Ocean to the Mediterranean Sea, east to the Pillars of Hercules, north along the western coastal plain of the Iberian Saudi Arabian Peninsula, north into Europe toward Scandinavia. They became cold climate adapted.

EUROPEAN GROUP

The northwest geographic area is West of the Ural Mountains in northern Russia parallel to 60 degree east in north longitude to the south end of the mountains at the Kazakhstan border, west to 50 degree at the Black Sea, and west to 47 and 37 degree in north longitude.

NOTE: Zero degree, east or west longitude from the north to the south poles. The zero degree division passes through London England, western France through the Pyrenee Mountains in Andorra, Spain, Algeria in Africa, and south into the Atlantic Ocean. These lines of longitude extend around the Earth to 180 degrees east and west longitude from pole-to-pole north-to-south.

Northern Europe: land area is 5 degrees west to 30 degrees east in north longitude and 55 to 70 degrees north latitude, northern Germany, northern European Russia, and Scandinavia; Iceland, Norway, Sweden, Finland, and Denmark.

British Isles: land area is 1 degree west to 13 degrees north longitude and 50 to 57 degrees north latitude, Scotland, United Kingdom, England, Ireland, and Wales.

Central Europe: land area 7 to 30 degrees east north longitude and 47 to 57 degrees north latitude; Netherlands, Germany, Poland, Belgim, Czechosolavia, Switzerland, Austria, Hungary, and Central Russia.

Southeast Europe: land area 15 to 25 degrees north longitude and 47 to 38 degrees north latitude, Romania, Yuguslavia, Bulgaria, Albania, Greece, and Eastern Russia.

Southern Europe: land area 40 to 60 degrees north longitude and 47 to 38 degrees north latitude south of the Black Sea; N. Turkey, N. Syria, N. Iraq, NE/NW. Iran, and south Central and south Eastern Russia.

France Spain, Portugal, Italy land area 3 west to 18 east degrees north longitude and 48 to 38 degrees north latitude.

The first groups of the migrations out of Africa 1.9-1.8 mya, were EH/LH, EG/EEG, and others variants. They took their tools and knowledge with them to the Mediterranean Sea in Egypt and the Middle East. They followed the shoreline north into Jordon, W. Syria, W. Turkey, and to the south shoreline of the Black Sea. It was here the group split into two migrations. One group migrated north to the Republic of Georgia (south of Central Russia), Shanidar, W. Iraq, and on to Teshik-Tash, Uzbekistan. The other group went west to Bulgaria and Croatia, where there was another split. EEG variant group migrated to Yugoslovenia, Italy, and to the English Channel, into France and south into Spain, pre-Neanderthal 1 mya-900 tya. It is not probable they crossed the English Channel at the straits of Dover into the British Isle. Other groups of HEB questioned claim EEG variant migrated north to Romania, Slovakia, Czech Republic and Germany. All of the migration was accomplished by 800 tya, when the climate was cold and wetter.

AS/ES variant group migrated to the Mediterranean Sea southeast shoreline in northern Africa. AS, LMS/EMS, HEB questioned claim EEG, migrated into the Middle East, Asia, Republic of Georgia, Uzbekistan, into Europe through Poland and Germany, west to the shore of the English Channel at Dover, and the British Isle becoming cold climate adapted 500 tya. They encountered the NEA living through out Europe 370 tya. Groups of HEB claim EEG and EAS variants migrated into eastern Asia and later migrated back into Europe through southern European Russia from north China, Poland, and Germany over land bridges 270 feet below sea level 70 tya.

A group of LNEA migrated to Belgium, Neander Valley Dusseldorf, Germany, Liberian Peninsula in Budapest, and Hungary 100-50 tya.

Advanced ADMS groups with increased global adaptability and nomadic traits migrated from the Nile Valley into Europe 45 tya.

Modern MCMS from France migrated into north and central Europe 50-40 tya.

MIDDLE EAST AND ASIAN GROUP

This part of Asia is east of the Ural Mountains in Central Russia at 60 degrees east longitude; the land area is 180 to 170 degrees east in north longitude and 70 to 10 degrees north latitude in south Central and Eastern Russia, Mongolia, and north China.

NOTE: At the equator zero degrees latitude is north or south in degrees. The land surface extends through central Indonesia; Sumatar, Borno, N. Sulawesi Islands, and into the the Pacific Ocean.

Middle East Region: south of the Black Sea and Caspian Sea at 27 to 50 degrees east in north longitude, and 47 degrees north latitude.

Southeast Asia land areas: Bangadesh, SW. China, E. India, Burma, SW. Thailand, and Malaysia; Indonesia Islands of Sumatra, Java, Sumbawa, Flores, Sulawesi, and Borneo.

Southwest Asia south of the Black Sea and Caspian Sea 30 to 75 degrees east in north longitude and 37 to 10 degrees north latitude; SE. Turkey, Iraq, E. Iran, Syria, Lebanon, Jordon, Israel, NW. Saudia Arebia, United Areb Emirants, W. Afghanistan, W. Pakistan, W. India, SW. Central and Eastern Russia, and NW. China.

The Middle East and Asia group: EH/LH, EEG/ADEG, and EE/ LE located at the Mediterranean Sea and from Egypt's Sinai Peninsula migrated to Jordan, Saudi Arabia, Iraq, Iran, Pakistan, southeast along the Persian Gulf and Arabian Sea, to the Bay of Bengal, India, Bangladesh, and into south China 1.15 mya-700 tya.

Late LE last migration thrust out of Africa was into the Middle East and Asia. They took their tools and knowledge with them 700 tya.

A group of ADEG settled in various areas in Middle East and Asia and became cold climate adapted. They migrated from China back to the Middle East, Europe, and along the south shoreline of the Mediterranean Sea, in Africa 600 tya. Some of the population each generation migrated a few miles further. Others stalled temporarily in the Middle East, Central Asia, northern China, and the southern part of Eastern Russia. It was very cold in Asia when AS, EMS and others migrated into the Middle East, Republic of Georgia, and Uzbekistan 500 tya.

European Asian LMS were transitioning to **Eurasian sapiens (EAS)** variants. This group migrated from central and south Europe east into the Middle East and Asia. EAS were not influenced by the Asian culture EEAPS.

EAS variants migrated into Asia, Kazakhstan, northern Sinkiang China, into northern Mongolia and the Sajan Mountains, to Lake Chatga, and to Lake Baikal in Eastern Russia where the group of EAS spilt. One group migrated north and followed the valleys and Lena River where trails of heard animals led them to Syaich, along the Verchojansku Mountain Range, near the Artic Circle where they became isolated 32 tya. Others may have split at the Lena and Aldedan rivers, south to Hot Springs where habitation was discovered 30 tya.

LMS EAS variants from the second group migrating east from Lake Baikal into Sichote-Alin, northern China and south Eastern Russia 25-15 tya. EAS became isolated for 10 thousand years. These EAS beget with the **Asian Chinese pithecanthropus sapiens** and transitioned as **LMS EAS East Asian Pl sapiens** (EEAPS).

Another group migrated back into Europe, because of the ice sheets, southwest through Central and west European Russia, Poland, and Germany over land bridges, when the sea level was below 270 feet 70 tya.

Another group split at Bangladesh along the Indian Ocean shoreline to Burma, Thailand, and Indonesia; Sumatra, Java and Bali where the, "migration of man", and other animals were prevented from crossing the Java Trench 1 mya.

NOTE: In the 1860's, Alfred Russell Wallace, established a theory about the, "migration of man", and animals crossing the Java Trench. An invisible biological barrier, known as the "Wallace Line", is located at the center of the Trench, between Bali, Indonesia, and Lombok Island. The 15 mile crossing of the Strait of Java Sea Trench prevented the Asians from negotiating or navigating treacherous currents flowing northeast from the Indian Ocean into the Java Sea.

LMS or EMS variants theory: may have discovered a way to successively navigate the Java Trench to Lombok Island and into Australia 60 tya. Lombok Island is a steamy volcanic island even today. This may be the answer to conflicting dated tools and stones used by Homo's, fashioned from the volcanic lava flow of hardened basalt. Later eruptions rained tons of ash over the landscape. This venture would have been less probable, and lacks Homo fossil discovery.

EE/LE/ADE, AS/ES/EMS (with GEO traits) migrated to Sumbawa Island, and settled along the Flores Chain of Islands where this group transitioned to H. floresiensis (FLO) and became isolated from further migration 95 tya.

Others may have migrated from the Greater Sunda Islands, Java Indonesia over linking land bridges when the sea level was 270 feet low, to Papua Guinea where human fossils were discovered.

Papua Guinea/New Guinea is 135 to 154 degrees east in south longitude and 5 to 10 degrees south latitude.

In New Guinea the seamount and coral reef islands of the Torres Strait land bridges would have provided crossing to Cape York Peninsula, Australia 75-70 tya, however, lacks the human fossil discovery.

AUSTRALIAN GROUP

Australian land area is south of Indonesia and the two Guinea's at 115 to 154 degrees east in south longitude and 10 to 45 degrees south latitude. The Territories are; Northern, Western, Queensland, New South Wales, Victoria, and the Great Desert Area.

Fossils of the **Indigent Aboriginals of Australia H. sapiens (IAAS)** variants were discovered along the coast line of Weber Basin north of Cape Crocker in the Northern Territory, Australia dated in stratum to 60-50 tya ca. EMS/AMS/IAAS robustus variants evolved at Cohuna, Australia 15 tya, and EMS/AMS/IAAS gracile variants evolved at Keilor, Australia 13 tya.

AMERICAN GROUP

The northern land area of North America is 170-55 degrees west in north longitude and 80-25 degrees north latitude. Canada land area is 140-50 degrees west in north longitude and 70-60-38 degrees north latitude. United States land area 170-65 degrees west in north longitude and 70-26 degrees north latitude. Contiguous 48 United States is 125-73 degrees west in north longitude and 48-43-46-25 degrees north latitude. Alaska is 165-140-130 degrees west in north longitude and 48-25 degrees north latitude. It includes the Aleutian Islands: 168 degrees east to 141 degrees west in north longitude, and 45-49 degrees north latitude.

Maximum glaciation was retreating north at 44 degrees north latitude in North America 20 tya.

Middle America (Central America): Mexico land area is 118-92 degrees west in north longitude and 32-15 degrees north latitude.

South Middle America land area is 93-78 degrees west in north longitude and 17-7 degrees north latitude through the countries of Belize, Guatemala, El Salvador, Honduras, Nicaragua, Costa Rica, and Panama.

South America

The land area is 82 degrees east to 35 degrees west in longitude and 10 degrees north to 65 degrees south latitude through the countries of

Venezuela, Colombia, Suriname, French Guiana, Brazil, Chile, Ecuador, Peru, Paraguay, Argentina, and Uruguay.

NOTE: The equator devides Earth laterally at zoro degrees starting point for north and south latitudes. In South America the equator zero degrees north and south latitude division passes through north Ecuador, south Colombia, north Brazil, and into the Atlantic Ocean. South latitude from the equator is south Ecuador, Peru, south Brazil, Bolivia, Chile, Paraguay, Argentina, and Uruguay.

The deglaciation and retreating of the ice sheet over Europe, Asia, and North America started 21 tya.

The Pacific Coast Migration Theory: This was a time when the sea level was below 400-300-200 feet. The glacier ice sheets opened over Beringia in Eastern Russia along the Pacific Ocean. Deglaciation of the Cordilleran Glacier Ice Sleet covering Western North Americas and the Pacific Coast was opening 21 tya. Some of these EEAPS variants were risk taking pioneers. About 18 tya EEAPS started to migrate north into Beringia, Siberia. EEAPS would have migrated from the China-Russian southern border at the Pacific Coast, north to Ostrov Island, Sakhalin, Russia, north along the coastal ridge crests of the Kuril Islands forming the Okhotsk Basin to the Sredinnyj Peninsula, north to Ploustrov Kamchatka, Beringia, Russia, along the ice covered Bering Sea Basin ice bridge into Central Alaska-Canadian Northwest Territory, south along the Pacific Ocean ice bridges and the Cordilleran Ice Glacier 17-16 tya. How they migrated along the Pacific Coast to the Americas', in cold climate, negotiating the Alaska-Canadian Northwest Territory, and Cordillean Glacier into a new world of North America is a continuing process of fact finding research. They would have followed the sea mammals and fish along Beringia-Alaska-Cordilleran Ice Glacier to the maximum southern glaciation and open land of North America, Washington State (USA) 15-12 tya.

EEAS were Western Paleo Americans below the ice sheet maximum in the United States on the west side of the Rocky Mountain Range.

Masses of ADE, ADMS, ADPI variants, and EEAPS were returning to Europe. They migrated from China after the Ice Age 65 tya, over land and ice bridges exposed when the sea level was below 400 feet. They crossed north of the Red Sea between the Horn of Africa and Arabia.

EAS/LMS/EEAPS variant group theory: some may have migrated from northern China along the Pacific coast south island-by-island when the sea level was below 400 feet into New Guinea, and into Australia 60 tya. But, again lacks the evidence.

Surviving EAS and EEAPS variants were both Paleo-Americans sapiens in North America below the glaciers and ice sheets in the USA. The Rocky Mountain Range and the Sierra Navada Range was covered with glacier ice extended into New Mexico, USA separated EEAPS on the west side and EAS on the east side 16 tya. They hunted the bison, mammoths and other animals of North America to survive 15 tya. North American EAS FNAS transitioned to Native American sapiens (NAS) 13-10 tya and later transitioned as Modern Native American sapiens (MNAS) in Canada, USA and Alaska, Central America, and South America. All that land was claimed by the NAS.

Chart 4.
Primate Population Location By Species

AFRICAN

MO=Morocca Sale AE
    Sidi Abderrahm AE
    Jebel Irhound AS EMS
    Temara Dar es Soitan Zouhrah Cave AE AS EMS
    Mugharet el Aliva AE AS EMS AMS
    Thomas Quarry AE AS
AL=Algeria Tighenif AE AS EMS LPI
    Taforalt AE AS EMS LPI
AN=Angola Chambuage Mine W. of Zambezi River AS
Cameroon
Mauretania
Mali
Nigeria
Mozamgique
Hwanda
    EG=Egypt   Nile River Valley ANEO

CHA=Chad   W Chad Djurab desert Lake Toros-Menalla
    BAH Bahagal el Ghazal Djurab desert Lake Chad SAT

SU=Sudan E Bir Ttarfawi AE AS
    Wadi Kubbaniya AS ES EMS AMS
    Singa AS
    Red Sea AS
ET=Ethiopia Gademotta AE
    Center Awash Valley
    Hadar/Gona Lake KEP ANA AFA Arr DRYA AE
    Adgantoli/Aramis #1&2 Indeterminate Amba Arr GAR
        E Bouri Modaj W Awash Basin Arrk
    Herto Bouri Peninsula Lava Dam Lake Yardi
    AE ADE AS
    Laetoli ABOI
    Konso Gardula AH AE AFA EE ERG BOIP
        AETP RUDCKERUD (RUDCE)
    Bodo LE AHEB* claim (RHO) EEG
    E Dire Dawa in E rift AS

ZI=Zimbabwe Bambata/Pomongwe N. of Limpopo River EH EEG AS EMS ADEG
ZM=Zambia (RH=Rhodesia) Kabwe Broken Hill Cave EH/ADEG AHEB* claim
(RHO) EEG
KN=Kenya N/NW. Lake Turkana KN/ET border; AFA AETP BOIP, Lake Ndutu
AHEB* claim (RHO) EEG
    Koru
    Cheboit Tungen Hills Lukeino Formation OT
    Heseloni PA/S Rusenga Island Lake Victoria CP PA
    W&E. Koobi Fora Lake Tukana (Rudolf) RUD
        ANA AH AE EEG AFRI AETP BOIP
        S Koobi E Allia Bay AHEB * claim ADEG AS FLB
    W Lothagam #1 Indeterminate AFA
    W Kanapoi S Lake Turkana AS EMS AMS
    Omo Lake Tukana AH AFA AE RUD BOI BOIP AETP AS EMS AMS

SA=South Africa Equus Cave Orange River FLB AS
    E. Florisbad Orange River FLB AS
    NW Elandsfontein FLB ES
    Orangia #1 S. branch Orange River FLB AS
    Klasies River Indian Ocean KLA AS
    Die Kelders Indian Ocean AS KLA
    Makapansgat SW of Limpopo River AFR2
    Harold's Bay Indian Ocean KLA AS
    Howieson's Poort Indian Occan KLA AS
    Cave of the Hearths AS

Blomos Cave Klasies River/Indian Ocean
    FLB KLA ES LFLB AMS
Saldanha Bay Atlantic Ocean AFR2 FLB ES ADE/EMS
    Hoedjiespunt (Hedijiespant)
    Hoodjipunt (Hedijies Punt) FLB ES AE GRA AFA ROB
    Langebaan Lagoon Peninsula AS EMS BOI others
    North West ASED ANAL AS
    S Geelbek AROB AS
*See text note for RHO

Djibouti
Niger
Ugonda

ZA=Zaire Matupi AS EMS

LI=Libya Central African Republic
    Ed Dabba AE AS
    Haji Ceiem AE AS
    Haua Fteah AE AS
TU=Tunisia Bir el Ater AE AS LPI
    El Guettar AE AS LPI
    Cape Horn Tunis AE AS
Burundi
TA=Tanzania E Chemeron S & KN border E Lake
    Victoria (Lake Baringo) # 2 Indeter AS
    N Serengeti Plain

    Chesowanja E Lake Victoria AE AFA
        BOIP/AOIP AS W Peninj BOIP
    Mumba W of Laetoli S Olduvai Gorge
        A/BOIP
    W Malema N Lake Malawi 500 miles S
    Uraha is S Malema at Lake Malawi ARUD
        BOIP AS
    W Olduvai Gorge Lake Rudolf ARUD AH
        AFR1 DRYA Lukenya Hill site and N
        Serengeti Plain

UG=Uganda SE Lake Victoria CP

Somalia

    KN Fort Ternen AS
        S&W Lake Turkana (Lake Rudolf) AE

NE. Kromdaai KN/TA line AH AFRI AETP
A/RUDP
Nariokotome
Krugersdorp

SA Florisbad FLB AHEB* claim ADEG AS ARUD/EH
NE Sterkfontein AFR2 AR/ROBP
Kromdaai AROB/ROBP
Swartkrans Cave AROB/ROBP AS
Drimolen Cave EH ROBP/BOIP
Bushman Rock AS
Border Cave AS

Johannesburg
Malopa Cave SED
Rising Star Cave NAL
Stekfontein
Burton
Taung Hills OR AFR2 AS
Witwatergerg

EUROPEAN

Iceland

Sweden

Denmark

SC=Scotland north central NEA

WA=Wales Pontnewyold ENEA AS ES
S. Goats Hole Cave on Grower Peninsula CNEA

England

SP=Spain Sierra de E Atapuerca Trinchera del Ferrocarril
Cave EEG AEC ADEG EEG EANT
ANT/NEA AS ES
Casa del Sol Zafarraya Andalusian Cave
CNEA/LNEA
La Sima de los Huesos Cave EHEB/EEG/EANT/ENEA
Zafarraya Gram's Cave/Forber Quarry EMS
LNEA/CNEA

FR=France La Rroche-a-Pierrot ENEA/LMS

FR Chapelle EMS

> Salon Noir MMMCMS
> Grotte Cosquer EMS
> S. Vezere Yelley LNEA
> N. Biesche St. Veast ENEA MMMCMS
> Engis La Naulette LNEA
> Chatelporron NEA
> La Charevet-Aux-Saints EMS ENEA
> Saint Cessaire LNEA
> Arago ES AS HEB (EEG)
> Ebingen DRYB
> Bretagne Celts and Brythons
> S. W. Cro-Magnon EMS MCMS
> Modern Magdelerthensis MCMS/MMMCMS
> Modern Aurignacian MCMS/MAMCMS

PO=Portugal Lagar Velho LNEA EMS/LNEA

Poland

BE=Belgium Biache ES AS LNEA
>   Spy Scladina Cave LNEA

GR=Germany Steinheim AS ES HEB EEG ENEA
>   Hanover EHEB EEG
>   Schoningen AHEB *claim/EHEB (EEG) EANT
>   Mauer NEA
>   Neander ENEA ADEG EHEB AS EANT
>   Heidelberg ENEA AS ADEG
>   Ehringsdorf ENEA
>   Vogelherd Cave
>   Neander ENEA

Austria

RO=Romania   Muierii EMS/LNEA
>   Pestera cu Oase Cave Carpathian Mt LNEA

IT=Italy Allamura ES AS
>   Ceprano EHEB/EEG
>   Saccopastore ENEA
>   Monte Ciruo ENEA

Liechtenstein

GE=Greece Pitrolona EEG/HEB/CNEA

CZ=Czechoslovakia Meadec EMS CAMS

CS=Czech-Slovic Dolnii Vestonice (Slovakia) EMS AS CAMS MMAMS

CZR=Czech Republic Pavlor EMS CAMS AMS
    (Czechia)   Kubna LNEA
                Ochog LNEA
                Sipka LNEA

IS=Israel   Qafzeh AS EMS
             Amud LNEA
             Kabara LNEA

Norway

Finland

Netherland

Ireland

BI=British Isles HEB/EEG

UK=United Kingdom Swanscombe AS ES
    Boxgrove AS ES EHEB/EEG
    Britton ENEA Celts and Brythons
    Hayonim ENEA
    Dover EHEB EEG ENEA
SP Cabezo Gorda AS ES
    Altamira EMS MMMCMS/FLB/KLA/ES
    Lesetxiki LNEA
    Gibraltar Gorham Cave EMS LNEA/CNEA AMS
    El Sidron NEA
        Pech del Aze
    Valles Penedes DRYL DRYC

FR La Quina EMS NEA LNEA
FR Lascaux Dordogne River Cave NEA

        MMCMS/ES/FLB/KLA
        La Ferrassie EMS LNEA
        Vegere Valley LNEA
        Combe Grenal LNEA
        Arcy-sur-Cure
        Mas-d Azil AMCMS

Niaux
St. Gaudens N. flank Pyrenies DRYF LNEA
Mariatal DRYB
S. N. SW. Atlantic Ocean MCMS LNEA
Le Moustier Mousterian Cave LNEA
Haute Garonne SI DRYF
Caune de Belvis LNEA/LMS

Luxemburg

Switzerland

GR Bitzingsieben AS ES EHEB/EEG EANT
      ADEG ENEA
    Thuringia
    Macur NEA
    Dusseldorf EHEB/EEG LANT ENEA
    Baden DRYB
    Wurttemburg DRYB
    Hohlenstein-Stadel

HU=Hungary   Erd LNEA
                Tata LNEA
                Rudabanya DRYB ANK

YU=Yugoslavia Petralona AS ES
                Muierii LNEA
                Pestera LNEA

Bulgaria

CR=Croatia   Krapina LNEA CAMS
                Vendija Cave LNEA

IS Skhul EMS AMS LNEA
    Tabum LNEA

ASIAN

Turkey

IRA=Iraq Shamidar Cave LNEA

SY=Syria   Zabzeh EMS AMS
           Zutteych LNEA

UZ=Uzbekistan   Teshik-Tash EMS LMS LNEA (EAS)

(W USSR)   S Sukhaya Mechetka EMS LMS LNEA (EAS)
(E USSR) Siberia SE Beringia LMS/PI (EAS & EEAS)
                    Ctr. Beringia Serpentine Hot Sptrigs EMS LMS (EAS)
                        S Altai Mt. Range Denisova Cave LMS LNEA (EAS)
                    N Verchojansku Mountain Range Artic Circle LMS EMS (EAS)

UKR=UKRAINE   W. Ctr. Mezhirich AMS (EAS)
                    Kiyih-Kaba LNEA (EAS)
                    Slarosillya LNEA (EAS)
                    Okladnikov LNEA (EAS)
                    Molodova LNEA (EAS)

RG=Republic of Georgia Dmanisi EH/EE GEO ENEA

South Korea

Mongolia

LI=Liberia Neckar ENEA, CNEA, LMS

Burma

TH=Thailand ANEO

Kampuchea (Kambodia)

IND=Indonesia Sumatera Lake Toba LNEA FLO
    Java Trinil Sole River LPI
    Bali LMS
        Ngandong HA/EEPI SOL ADE/ADPI
        Yogyakarta ADPI
        Jakarta ADPI
        Mojokerto ADPI
        Sangiam Basin ADPI

NK=North Korea JIN ADE/EPI

NG=New Guinea LMS AMS EEAPS
AU=Australia Weber Basin IAAS LMS
        Cape Crocker IAAS LMS AMS
        Lake Mumgo Willandra Sys
        Jinmium IAAS LMS
        Cohuma IAAS AMS

Iran

Afganistan

Pakistan

IND=India Narmada EEG EH/EE
      S Arabian Peninsula EHEB/EEG
      Naroda EEG EE

CH=China ANEO
  Zhaukaudian Cave EE EPI
  Dingcum ASPI
  Changyang ASPI
  Tongzi ASPI
  Maba ASPI
  Nihewan ASPI
  Yiyuan ASPI
  Nanjing ASPI ·
  Herian ASPI
  Youonmou ASPI

CH=China Dali ASPI
  Yunxian Province Hubei EPI/LE
  Beijing SI AEG EPI AEG/AE/API ASPI
  NE Sichote-Alin LNEA LMS/PI (EEAS)

Japan

Vietnam

Taiwan

Malaya

IND Sumbawa Island FE FLO LMS PI
   Borneo Niah Cave H. sapiens
   Timor Island Jerimalai EE EMS
   Sulawesi
   Lombok Island EE EMS
   Flores Chain Islands EE EPI FLO
    Soa Basin FLO EPI

MY=Myanmar ANEO

AU Malakunanja IAAS LMS AMS
   Nuwalabia IAAS LMS AMS
   Keilor IAAS AMS

AMERICAS'

North America

Canada
    Western Cordilleran/Eastern Laurentide Glaciers
    Rocky Mt. Alberta Canadian Passageway FNAS
        Charlie Lake Cave FNAS
    East and north Atlantic Ocean coast, five Great Lakes, FNAS, Norse, MNAS,
        French, British, European settlers

United States AL=Alaska Yukon Northwrst Territories
    Tuluaz LMS FNAS
    Nogahaabara LMS FNAS
    Old Crow LMS FNAS
    Bluefish Caverns LMS FNAS
    Swan Point LMS FNAS
Alaska S Pacific Ocean Coast EMS LMS PI (EEAPS) (EPAS)
Western Montana Anzick FNAS
Schaefer/Hebior, IL FNAS (WPAS)
Meadowcroft, OH FNAS (WPAS)
Page Ladson, FL FNAS (WPAS)
Linsay, ID FNAS (WPAS)
Debra L. Franklin Austin, TX FNAS (WPAS)
Horn Shelter, TX EEAPS (EPAS)
Palsley Caves Columbia, Klamath River, OR FNAS (WPAS)
Columbia River LMS EAS FNAS Ancient One, reburial to ANAS: CT of the UIR,
    B of the YN, NPT, & WB of PR
Channal Islands, CA EEAPS (EPAS)
Manis, WA EEAPS (EPAS)
Richland, WA LMS EAS (Caucasion?), (FNAS?)
Northern US Montana to the Atlantic Ocean LMS EAS FNAS (WPAS), MNAS,
ANAS
East and south coast of the Atlantic Ocean NAS, MNAS

Middle America
    Mexico (EAS) NAS
    Yucatan
South American
    Monte Verde, Chile FNAS
    Arroyo Seco 2 Sumidouro Cave, Brazil FNAS

TOOL TECHNOLOGY AND
CULTURE EVENTS

There were no makers of stone tools discovered 3 mya. The first primitive stone tools appeared about 2.6 mya.

More primitive spears were sharpened sticks. The wooden hand held bayonet was sharpened on one end and the other end was formed into a bulb to assist in thrusting into the prey at animal kills. The bulb alternate use was for grubbing in the dirt for tuber root plants. There was high risk of harm to these hominids who encountered injury from animal hunting to their hands, head, arms and upper tarsal.

Bone tools were used for digging 1.8-1 mya. Animal bones were found that show heavy wear on the tips and parallel along the shaft.

Stone Age tools were used 2.6-2.3-300 mya and as late as 10-4 tya. Olduvai tool technology discoveries were part of the Pleocene Epach 2.6-1.8 mya. The tools continued to be used into the Plestocene Epach 1.8-1.5 mya, overlapping new Acheulean technology 1.5 mya-100 tya, overlapping into the Holocene Epach to 40 tya, and the Acheulean-Uberdiya tool technique overlaped Acheulean tool technology 1.5 mya. Later Acheulean tools continued to be used through 1.5 mya-100 tya. Acheulean-Sangiran technique 600 tya was used until 200 tya. Acheulean-Neolithic technique was used 650-500 tya. Acheulean-Zhoukoudian technique and the Acheulean-Levalloian techniques were developed in France 300 tya. Acheulean-Kapthurin technique is divided into the Lower Palaeolithic and Middle Palaeolithic techniques in Europe and Acheulean-Mousterian technique about 150 tya. The Acheulean-Upper Palaeolithic technique was used about 50 tya. The last Mesolithic technique started about 10-4 tya.

Blade tools were used widespread in Europe 400 tya. Asia 400-150-50 tya, and Africa 300-250 tya.

At the Pennacle Point Cave, South African culture used red ochre art, painting themselves, and cave walls with their visions of spirts and animals 164 tya.

Some of the bones from animals preserved in the stratum displayed cut marks from the tools.

Olduvai Stone Age Technology tools were the oldest, consisting of stone; simple pebbles, hammers, chopping, and sharp edged flakes from the river beds.

Australopithecus early hominids learned to select two hand size stones. One was used as a hammer stone and the other was used as the core stone. The hammer stone struct the core stone at an angle breaking flakes. The flakes were used as cutting tools. The core was

shaped for the hand grip, with sharp edges and was used as a chopping tool. These kind of tools were discovered at butchery sites, among animal bones, littering the area 1.9-1.2 mya. These tools overlaped newer stone tool techniques in use until 1.83-1.53 mya.

Tools were discovered in Africa; Awash Valley Region at Omo, Hadar/ Gona, Senga Ethiopia, and Lokalei, Kenya, dated to 2.6-2.3 mya. Also, tools were discovered in Lantian, Asia 1.9 mya, Sterkfontein South Africa 1.8-1 mya, and in northern Spain in Atapuerca, 1.6 mya.

Olduvei-Uberdiya technique was the start of the fabrication of a new handax and the Acheulean Technology. This technique was used and overlapped Acheulean until about 200 tya. Select flat oval or pear shaped stones from the river beds were fashened to a sharp edged point. The other end was rounded to fit the palm of the hand. These handaxes were used in northeast and middle Africa, at Dmanisi and Riwat in Asia 1.5 mya-200 tya.

In the Awash region, at Bodo, Ethiopia, Africa, stone hand axes and cleavers of the Acheulean technique were discovered in stratum dated to 800-600 tya ca.

The new multi purpose handax tool had three different sides; a crushing hammer, cutting edge, and a scraping surface for butchering with a palm smooth surface between each tool surface for the palm and fingers to grip during the tool process of larger animals 1.5 mya-40 tya. Bones of antelope, horses, and other animals exhibited tell tale cut marks on the surfaces from stone tools.

The **Acheulean Stone Age Technology tools**: ADE developed Acheulean Stone Age tool technique with the knowledge of mentally mapping the strike "knapping" a basalt core rock at the correct angle to produce sharp flacks on all sides. They crafted tear drop shaped spear heads. ADE crafted symmetrical tri-side **hand stone axes** to fit snuggly in the palm of the hand. The way it was held could cut, crush, or batter, animal skin and bones 1.5 mya-40 tya. There were many versions and techniques used as survival tools were developed. Handax, variations of new types of arrowhead shaped points and serrated harpoon like points were discovered in northwest, central, northeast Africa, Bodo, Ethiopia, South Africa dated to 800-600 tya, in Parkfield eastern England, and Asia 700 tya.

Later Acheulean tools were used; bifacial choppers with flukes on two sides with sharp edges providing a more efficient ax like chopper, scrapper, pointed awl, fashened as a symmetrical tri-side disc shaped polyhedral tool for butchering with a palm smooth surface between each tool surface for the palm and fingers to grip during the tool process of larger animals. Some of the bones from animals preserved in the stratum displayed cut marks from the tools. This tool processing took a relatively high order of intelligence and dexterity to accomplish 500-100 tya.

Fire was used 400 tya. At Swartkrams, South Africa they fire hardened wood spear tips.

Acueulean-Neolithic stone tool technique was discovered and used 650-500 tya.

Acheulean-Sangiran stone tool technique was discovered and used 600 tya.

Acheulean-Zhaukaudian stone tool technique was discovered and used in Eurasia 500 tya, Nile River Valley Egypt, Africa, Swanscombe and Kent in south England, Boxgrove, and Hoxne in east England (Briton) 400 tya.

Near Hanover, Germany, at the Schoningen butchery site near Hanover, Germany, seven wooden glistening dark brown hunting spears 6 feet 6 inches long were discovered in a peat bog strip mine preserved by the tannic acid were dated to 400-300 tya. The spears were balance weighted and were probably throwing spears which reduced the personal damage risk at the hunts.

At the Bilzingsleben butchery site in eastern Germany: a verity of artifacts and 200,000 stone tools; hand axes, hand held three sided cutting-scraping-crushing hammer stones were used until 250 tya; 100 large and small animal bone; wood and antlers; 37 H. heidelbergensis (HEB) claim were H. ergaster and H. neanderthalensis variant species fossils; and 36 plant species all preserved in calcium carbonate 400-380 tya.

Acheulean-Levalloia stone tool technique was discovered and used in France 300 tya.

Acheulean-Kaphurin stone tool technique was split; into the **Middle Palaeolithic technique** stone blade used for cutting flesh and wood 250 tya, and the **Lower Palaeolithic technique** stone tips were used for wooden thrusting spears. The points after use could be modified by resharpening the edges and were used 200 tya.

Acheulean-Mousterian stone tool technique was discovered in Europe. Large flakes were struck from a core stone used to scraping hides 150 tya.

Acheulean Upper Paleolithic stone blade tool technique was used widespread in Europe and Asia 74-50 tya.

The **Mousterian Period** started 125 tya, in Europe. It overlapped the Acheulean Stone Age ending 4 tya. From a stone core, struck with a single blow, sharp flakes "Levallois Point", were discovered at a French site. After it was in use and the edges became worn the tool was restored by sharpening the points to sharp edges.

The **Mesolithic Tool Age** was the end of the Stone Age technology producing composits and microlithics, thin blades, bone, ivory, and antler tools in Europe 10 tya. The Achrulean technology overlapped into the Holocene Epach to 4 tya.

The **Aegean** (sea between Greece and the Asian Minor) cultural and technology changed. It was the start of the early Bronze Age, 28 tya.

CULTURAL PERIOD

H. Georgicus with AH/EH/EE traits lived togather 1.8-1.77-1.6-1 mya. The **Acheulean Cultural Period** started 500 tya.

The **Flo Culture** population evolved as hobbit dwarfing on Flores Island, Indonesia 95-18 tya.

The **Magdalenian Period of Culture** was discovered in Salon Noir, France 50 tya.

The **Chatelperronian Period of Culture** started in France 45 tya.

The **Cro-Magnun Culture of Modern H. sapiens** in Europe, Middle Asia, and Africa dated to 40-28 tya.

The **Aurignacian Period of Culture** and populations of modern humans colonized along the Mediterranean coast in Italy, France, Spain, and the Black Sea Danube 50-45 tya. Technology and art was developed; figurative carvings, musical instruments were fabricated in France 30-28 tya. Carvings were discovered of mystical lion-man were spiritually symbolic from Hohlenstein-Sadel, Germany 30 tya. In a Voglherd Cave in Germany a bone flute was discovered dated to 35 tya.

The **Gravettean World Age Period of Culture** and new tool technology was developed by the Cro-Magnon, modern man, 32-25 tya in France.

The **Neolithic Period of Culture** and technology was developed in Predmosti, Czech Republic. They fabricated bone needles and sewing of clothing 28-24 tya. In Mezhirich, Ukraine wool and skin structure shelters were constructed 17-12 tya.

The **Moravian Culture** and technology was developed in Czech-Slovak area of Asia, 20 tya.

The **NEA Uluzzian Culture** and technology was discovered in Italy and Greece 20 tya. Their tools were improvements after the advance of modern humans.

PART 2

AFRICAN EUROPEAN VIOLANCE AND BEHAVIOR

Homo sapiens with their advanced mental ability learned aggressive behavior toward others for self gain. This behavior developed into violance in the form of bloodshead of other lesser combative individuals, death, and often cannibalization for survival. As homo sapiens became more violant, some banded together. These uncivilized barbarians lived a life of pillage, lacked moral care, and any concept of good or evil. There was conflict in how others treated each other that lead to war. There needed to be more communication, understanding, education, and love between others. Some Homo sapiens still have not learned how to control violence and behavior.

Violance is the use of force or physical compulsion to abuse or damage.

Behavior is holding oneself in a certain way or manner the individual acts.

Civilization are changes in the style of life; social differences, appearance of class and categories, and mental health.

Confrontation management is meeting and negotiating an agreement; a plan of agreed corrective action, implement the plan agreed upon, follow up at later time, and make any corrections agreed upon. If there is a disagreement, terminate the meeting, and try again later with new information.

Education; anger management, emotional control, and stress management.

Love; is having affection, as humans, for other humans.

Pagan; is a person having no religion.

Belief; conviction; acceptance of a creed and doctrine as an individual **faith**; unquestioned belief in anything believed as a **Creed**; belief in religious stated **Doctrine**; something taught as the principals of **Religion**: any specific system of belief conduct and ritual involving a code of ethics.

God; any of various kinds of beings that have existed; conceived, supernatural, and immortal with special power over humans **Supernatural**; existes or occurs outside of known forces of a natural **Immortal**; a person having lasting fame **Natural**; qualities that make something what it is, the essence; an entity; as being; existence.

Violance and behavior issues were not in the evolutionary legacy or in the human genes. The learned aggression and defensive attitude escallation dominated in a struggle to survive and for power over others. H. neanderthalensis, "Neanderthal", cousin to H. Modern sapiens, lived in the climate that was extreme and rapidly fluctuating 60-40 tya. Most H. sapiens did not tolerate the challenged NEA in social interaction. NEA and ADMS were probably relatively swift and hostile excluding each other.

More violence and behavior aggression developed by the male species with high testosterone hormone characteristics lacked the education needed to control their emotional stress. Pilfering, taking what was not theirs from those who have, and bloodlust with vengeance was common in those Dark Ages of time. Change was needed to decrease physical violence, mental health, and non-evasive straightforward crime. Mixed European ancestry culture was the result of unrest and wars. Many survivers sailed in many ships hulls to a new world experience. They learned to adapt to foreign British, French and Native American sapiens in the quest for land on the Atlantic coast, of what is now the United States of America.

For the common cause, civilization is to reach a high state of social and cultural development without ill behavior.

Civilization or the lack of it, as the population grew. Social differentations and appearances of class and categories were assessed. Priests, soldiers, artisans, and aristocrats contributed nothing to commerce, feed by the peasents. This satisfaction resulted in ownership and exploitation of land. The peasants developed Agriculture and Technology and had to compete with hunter gather economy. Changes in the styles of life from irrigation farming or dry agriculture increased demands on the land. Non-human sources of energy was used by animal power; oxen, raindeer or later horses, used in plowing and primitive tools to cultivate fields. Tool technology inventions made the agriculture tasks more efficient.

AFRICAN SOCIAL CULTURE
AND BEHAVIOR

Homo population grew along the Nile River, Africa, including those who stalled from the Eurasian EAS migrations back to Africa. In the Nile River Valley of Egypt, at a site called Wadi Kubbaniya, conflict arose among the ADMS and other inhabitants 22-20 tya. There were confrontations and competition within the species to survive. Learned ill behavior was caused by the inability to control emotion and stress that resulted in violance, fighting, and killing over the scarcity of resources. In Sudan massave grave sites were discovered dating to 14 tya. The killings by each other were with spears and clubs. It looked like systematic warfare and cannibalization to survive. Women and children were the venerable ones.

The War of Mallin in Kenya was 9000 years ago. AMS killed each other when they started to compete for power. Once AMS was established as the dominate specie throughout the region they systematically sentenced the enemies with influence and thoses of the opposition to death, including the misfits and ill behaved, and those who could not adapt or change 10-8 tya.

EUROPEAN ASIAN SOCIAL CULTURE
AND BEHAVIOR

Central Europe ancestors HEB, ADEG, NEA, AS, ES, MS, AMS, and EAS created the European species of multiple mixed variants. Many shared H. sapiens ancestory of peace, good, bad, and evil.

AMS applied simple **human behavior** concepts. This was a civilization who shared in team work and social culture. They developed their technology; bone tools, needles to sew animal hides into weather resistant clothing they wore, crafted woven blankets and rugs from textiles, fabricated fishing and animal nets 300 feet long and 3 feet high, fashioned figurines using stone tools, and used bone musical instruments 17-14 tya. They had a **creative** sense of prospective; crafting ceramic artwork and coatings, expressing themselves; in pictures, painting, and carving of objects in search of meaning in social culture. They discovered high heat caused the sand (silica) to turn into glass below their hearths used in their art. There were elaborate funeral practices performed for tribal social hierarchies. Their penal system judged those with bad behavior or could not adapt to social culture. Some were not tolerated and some were exiled into ever shrinking pockets of habitation, namely NEA.

Some resorted to cannibalization to survive. Some failed to survive adapting to their environment.

In the village of Mezhirich, Ukraine, AMS shelters were built of mammoth tusks and bone in round foundations. Pits were dug in the permafrost to freeze their meat from the hunts. They may have been encouraged to settle giving up nomad mobility and living by following the rules of the tribe 15 tya.

Competition among the world's humans changed radically with the development of agriculture 10 to 5 tya. Some of the plants and animals were domesticated, which meant they could grow there own food and not have to rely on cooperative hunts or seasonal gatherings from the wild. Permanent villages increased their population and fed the specialists; inventors, solders, and kings who did not produce food. Technology advanced to metal tools, writing and state societies. The first farming was in southwestern Asia; wheat, barley, sheep, cattle, and goats were domesticated. These Pagans practiced ritual based on the sun, moon, and Earth belief.

There were other AMS variants who contributed significantly to family biological adaptation in their environments evolving unique trait changes in life, form, and social cultures. They were learning more about the Spirit World fears and beliefs.

H. Moravian sapiens (MAMS) variants were tribal **Eurasians** (EAS) living in northern Asia at Czeck-Slavic sites in civilized communities. They were a civilization who were cooking, heating with hot rocks, creating art objects from ceramics; clay pottery, animal and human figurines, weaving; bags, blankets, clothing, and hunting nets fabricated to capture small game. There were funereal practices with ritual burials. The Pagan belief was with the sun, moon, and Earth. All their tasks, for the most, were accomplished by cooperative team work 20 tya.

Many of the Europeans were of mixed variant ancestry. They were the same population who were transported in merchant ships, crossing the Atlantic Ocean in about one month, to a new land surviving violance, settled in colonies in North America that became the United States of America.

How The British Isles Began In Europe

The Celts and Brythons from Northern Europe migrated to the western territory of Welshland (later Wales), the southern territory along the English Channel (later south Briton) and Bretagne, France long before the Ice Age 23 tya. There were Celts and Brython populations living in tribes in Ireland, and Welshland (later Wales). The Scots from Scotland, "Caledonians", migrated to the south territory (later southern Briton). All of these migrations were before the Roman occupation.

The Roman occupation (later Briton) was in 43 AD. They spoke Latin. Romans were trained legionairs and were skilled in arms. The Celts were descendents of Welsh nobility under Roman rule. They spoke a language of Brythonic of Celtic origin and Pict. They were Pagans when the Spirit concept of Christian values was introduced in 100 AD. The Celts cavalry resisted Roman incursions using outdated chariots 175 AD.

Roman forces inlisted European warriers from Sermatia, north of the Black Sea (in now Ukrain) into Roman Service. Their cavalrymen wore scaled armor composits of metal "cataphracts". This armor was heavy and less flexible than chain mail. They were armed with a lance on a horse and a sword or dagger as a backup weapon 200 AD. The light weight sword "seax" (say-aks) was used with one hand. Also, it was used as a hunting knife or dagger. It was single edged and more versatile than a full sword.

The later tribes were loyal to Roman rule for 350 years until the Romans withdrew in 410 AD. The dominate language was Brythonic-Celt.

The Vikings in Europe

The name "**Viking**" comes from the Norse language and means "piracy".

The Danish built warships were open hull, with overlapping planks on the deck matched to the bow from stem to stem. Oak was used in the classic Scandinavian forms with tall masts equipped with multi-sails powered by the wind or maned by the crew using herring bone array oars. There was very little deck housing or cabins exposed to the harsh cold elements. Small fires were built in special sand box areas below the deck of the ship's hull.

Norseman Vikings sailed the North Sea, the English Channel, and the Atlantic Ocean coast in strong Danish warships propelled by the wind or by manpower using oars. They were protected from the cold clothed in their animal skins.

In the years 400-600 AD the Jutes from Denmark, the Angles and Saxons from northern Germanic tribes sailed the North Sea in their Danish warships to the territory (later Briton) east coast.

The Scots were from Ireland and the Picts were from northern territory (later northern British Isle). They invaded the west and southwest central territories.

The Jutes were Scandinavian Vikings. They sailed their Danish warships in the North Sea to the southeastern territory (later Briton), along the northern English Channel, and battled with the Roman-Briton forces as they pilfered the land and population.

The Angles Germanic Vikings sailed their Danish warships in the North Sea to the east coast, central, and northern territories (later Briton) to pilfer the land, population, and to battle with the Roman-Briton forces. Later they allied with the Roman-Briton forces and pushedback the Picts to their northern controlled territory (later northern British Isle). Angles became allies with the Saxons and helped the Roman-Briton forces pushed back the Scots from the central and south territories (later Briton) to Welshland territory west central land (later Wales).

The Saxons Germanic Vikings sailed their Danish warships in the North Sea to the east coast, central, and southern territories (later Briton), battled the Roman-Briton forces, pilfering the land, and population. They allied with the Roman-Briton forces and helped pushback the Scots to what was Welshland territory (later Wales) and forced the Picts to retreated back to their northern territory at Hadian's Wall (73 miles across northern Briton). The wall was the boarder dividing, southern Kingdoms from Pict Territory (later northern British Isle), where the battle of Heavenfield took place in 634 AD.

The Scots sailed the Atlantic Ocean coast, the English Channel, and to Normandy, France.

The Britons were colonizers and ruled for 350 years by the Romans. The Romans withdrew their forces from Briton in 410 AD. The Jutes, Angles, and Saxons invaded Briton and the population. The Britons were poorly equipped to resist the Saxons after the Romans withdrew. The population became Anglo-Saxon with unquestioned obedience under authoritarian rule. Territories were colonized into three Kingdoms; Wessex Kingdom in the south, Mercia Kingdom in the central land, and Northumbria Kingdom in the north to Edinburgh 450 AD. Hadian's Wall was 96 miles, east to west, dividing the Northumbria Kingdom northern border and the Pict Territory. The west central land was Welshland boardering the Mercia Kingdom to the east. The Saxon King of Wessex defeated Brythonic forces at Gloucestershire, Somerset, and Oxfortshire (later England). The Brythonic language was gradually replaced by Anglo-Saxon Old English. However, the language survived in Brittany, Wales, and Cornwall.

The ledgendary King Arthur was of Roman-Briton variant nobility, robinhood of Welshland, 400-450 AD, a relative of a barbarian prince Ambrosium Aurelianus, with his band of Knights of the Roundtable, in Welshland territory, late 575 to early 600 AD, and was often called King Arthur. Britons Army commander Ambrosius Aurelianus was of Roman-Briton variant nobility from Welshland Territory. In the Saxon language, "cniht" was a person of nobility who did not see themself as knights or know anything about chivalry. Britons battled Scots at Badon Hill, Sumerset, Witshire, south Scotland, and various points along the southern coast of the English Channel some time in late 500 AD.

Britons were ruled by the Saxons in the early 600's AD. Pagan Germanic values dominated. They built fortifications and cathedrals for social control to keep the peasantry in line at Canterbury and York 700 AD. Initial type of fortification was a main structure on a mound with a surrounding ditch (motte), an outer area was protected by vertical wooden poles (bailey), and another ditch offering limited protection. Stone castles begain to be built and stone walls appeared around towns in Northumbria Kingdom at Alnwick 1100 AD. This type of fort made surprise attacts less likely. The "Coat of Arms" was a coat of metal plates attached to a surcoat.

Mercians were Angle Vikings, "Mierce" from the Old English meaning "frontier people". They ruled for 600 years. One of the 18 wars was the Battle of Heavenfield 634 AD with the Northumbria Kingdom. They were boarder raiders between Mercia and Welshland territory in the Staffordshieres Trent Valley near Lichfield in 650 AD. One of 14 wars was the Battle at Hastings at Sussex with the Wessex Kingdom, and 11 wars with the Welsh Kingdom. Angle-Saxons variants from the Mercia Kingdom authority extended to the northern Pict Territory, Scotland, Ireland, and Welshland 600-850 AD.

Viking sea raiders from Norway, Sweden, and Denmark tribes were known as the "Danes". They invaded the east coast (later Great Britain) 790 AD. On June 8th 793 AD these Pagans stormed the monastery of Lindisfarne in Northumbria at the North Sea. Norseman invaded the English Channel and took over the Angle-Saxon Kingdoms on the British Isle 793 AD. In 875 AD Saxon King of Wessex Alfred the Great 871-899 AD was the only Kingdom remaining. Alfred's grandson Eadred ruled a United England 920-955 AD.

The Viking marauders sailed into the English Channel surging through the British Isle, Baltic States, and northwestern France during the 9th century AD. Charles the Simple, King of France, creed Normandy to a Viking chieftain named Rollo, who protected the coast against avaricious Norsemen. The Vikings commandeered East Anglia of the Mercia Kingdom, city of York and was the King Jorvic Viking Center in 865 AD. The Viking King of York Erik Bloodaxe was deposed in 954 AD. The Anglo-Saxon and Normans ruled the territory of the English Channel to the border of Scotland during the Kings of Europe. The two ruthless kings Harold Hardrada 1st of Norway and Duke "William the Conquer" of Normandy were descendants of Rollo and related to Nordic ancestry and Briton royalty. Vikings were raiding monasteries for the money, silver, and gold in 911 AD. These Nordic Viking "Dane" raiders pushed through Scotland, Ireland, the eastern one-third of England, and northwestern France.

The part of France closest to England is Normandy. Normans commanded by Duke William crossed the English Channel, invaded England in the Wessex Kingdom at Sussex at the Battle of Hastings, known as "The Norman Conquest". Normans and Harold 1st, archery, infrantry, and cavalry attacks defeated the English army 1066 AD. William's conquerst of the three Angle-Saxon Kindoms paved the road to William's claiming the English throne. William spoke Norman French and disliked the language of the Norse established in England. The English language was indelibly marked with words derived from Icelandic origin.

The Scandinavian Danes were inhabited by tribes who spoke Norse, a Germanic language 1000 AD. Hardrada's army was pushed back at the Battle of Stamford Bridge, near York, by the English Army. Their about face confronted the Normans who had sailed across the English Channel. Together the Vikings overrun the English. The armies of Scandinavians Danes held the throne. Harold II was crowned king. The territories ruled the urbanites for 200 years, 1000-1200 AD. Large parts of eastern Briton extended to the border of Scotland, known as the lands of Dane Law, and included the northwest Isle of Man, 1100 AD. There were mixed populations living togather (later British Isles). William systematically deported misfit and criminals, aboard merchant ships to Australia.

Pagans gathered at the Stonehenge Center, near Wittshire, England each summer solstice when the sun ecliptic is at the farthest north of Earths equator June 21 or 22 of each year. They practiced ritual ceremonies of human offerings.

Norwegian and Danish Vikings colonized the Orkneys, Shetlands, and Hebrides. They harassed Wales and the Northwest Isle of Man, another staging area for Viking raiders. In South Wales during the **Monastic Period**, Kings ruled from Roman fortress 1100 AD. Knights, war armor, equipment, and iron swords were from the Iron Age 1100 AD.

The elite of the army were mounted on a horse with a lance, sword, and war armor over a chain mail; flexible body armor links of small metal rings, scales or loops of chain. The added weight was tiring. It became a problem for those who fell surrounded by enemies. The ground forces used specialized armor piercing war hammers, picks, and heavy axes.

War armor consisted of metal attached to leather. The "helm" (helmet), was metal padded with horse hair or wool with cheek plates. Later war armor improvements: Great Helm; covered the entite head reducing visibility and ventilation. It was worn over a chain mail coif or padded hood under it, adding a small metal helmet, "bascinet", fitted inside the Great Helm 1300 AD. A hinged visor was added to the helm, and each were of different fabrications up to 1450 AD. The "Suit of Armor" consisted of; "Gorget" around the neck, "Pauldron" sholder plates, Lance rest for the right arm, "Gauntlets" arm and hand gloves, breast and back plates, "Tasse" waist and thigh skirts, Leg coverings; upper "Cuisse", Kneepiece, lower "Greve", and "Sollerets" for the feet.

In the late Medieval times Renaissance did not reach Great Britian until Queen Elizabeth 1st reign 1533-1603 AD, associated with the increased use of firearms.

The Hand-gonnes Cannon was a two man task. The metal pot below a long pipe was loaded with nitrate slow burn powder. A soft metal ball was inserted into the interior of the pipe at one end. That end was inserted through a fitted hole in the side of the pot and locked down in place. Two men held the cannon aimed at a target. The powder in the pot was ignited by a third man. The gas pressure built up in the pot and propelled the ball out of the pipe. It was not very accurate.

The Teutons in Europe

As the social culture changed, or the lack of it, these Icelanders, Norseman, Danes; from Norway, Swedon, and Denmark, "Nordic raiders", were uncivilized barbarians who invaded northern Europe and central Europe with a vengence to secure power, wealth, and dominate others.

These Teutons raided gained wealth in seasonal raids along the European eastern Atlantic Ocean coastlines. Pilage was a way of life, free of moral care or value. These tribes were free people acting on their own authority and honor. They gathered their followers in search for revenge and the spoils left for their taking. They lacked absolute concepts of good or evil. Their virtues were of honor, battlefield valor, and they were all chiefs. They held slaves and exalted bloodlust in the competition for power among the tribes. When the Vikings consolidated with others, they forced the rival chieftain into exile.

Teutonics tribes invaded Central Europe. They were from northern European Germanic and Scandenavian territories. They were Goths, Visigoths, Ostrogoths, Huns and Vandals, the last two were uncivilized and ill behaved barbarous Teutons 1000 BC through 500 AD. The Goths ravaged and pilfered the land 150-200 AD. The Vandals occupied from 170-548 AD. The Ostrogoths occupied from 200-454 AD. They were the bearded warriers carrying their iron swords into battle, wearing woolen clothing 350 AD. Huns occupied 375-425 AD. The Visigoths occupied 376-418 AD. They established their outposts of their own kind (natural group). They created hate for strangers and foreigners establishing their own warrior code.

The Danish built warships were used by these Teutonics. Nordic barbaric raiders along the European Atlantic coast pilfered through out the land. They were the same groups of Icelanders, Danes, and Norseman Vikings who migrated into Europe 100 through 1000 AD. Teutonic military personnel were equipped with long cutting blade swords about 3 feet long. The sword fabrication technique; twisting rods and strips of iron forged and hammered togather making the steel blade, forged again and hammered into dubble cutting edges, and finish by polishing.

The Roman Empire region was systematically invaded by Viking-Teutons. They took over the Roman Empire. They established their outposts of their own kind, adopted new lifestyles, cultures, customs, and were Pagans.

Central Europe was divided and vulnerable after centuries of tribal invasions that followed the collapse of the Western Roman Empire in 476 AD.

The Valkyries were lusty maidens who swept death over central Europe and were the heroes to Valhalla.

The Swedes invaded Russia along rivers to Constantenople and the Orient.

Magyars from Italy, Hungry, and Yugoslavia, invaded southern Eurasia. The Slavs bordered the east side of the former Western Roman Empire in 814 AD.

The Magyars occupied the former eastern Roman Empire and all of The Carolingian Empire as the Feudalism, formally the Prefecture's of Gaul and Italy. They claimed the land and its people as their subjects with demands for loyalty to them or they were systematically disposed of. The Magyars conquered the Moravian Empire, establishing Hungary in 904-1848 AD.

The Magyars and the Norseman Viking's grudgingly became civilized, accepted Christianity of the spirit world concept. They learned from stories of the spirit world encouragement that became their belief and faith, usually by kings or leaders of tribes. It was the time of central power of kings in 1000 AD. They adopted new lifestyles, cultures, and customs.

The over population of H. sapiens variants became too numerous to keep account of. Only the dominate leaders of tribes had one surname. The followers were no-named people of the leader. They were loyal to their leaders, but their relationship or ancestry could not be established. Kings needed to identify their peasants for food and tax purposes. The churches scollars were educated and could read and write. They were deligated by the King to keep records and assign names to individuals for tax purposes. Most of these now named people had a language, but, could not read or write.

The population was organized into kingdoms, with churches granted vast properties for their agreement of record keeping and tax collection. The churches had reciprocal obligations using Vassals and Lords. The churches begin to form the primary social organization from local manor to kingdoms. There was some regions independent of the city-state federated in leagues. The mixed species variant of populations grew. They developed agriculture, had better communications, towns were developed, and trade was established. The population was not educated to read or write.

North and South Americas'

EAS migrated from Eastern Russia Bering Strait ice and land bridges connected to the Alaska-Canadian Northwest Territory. They lived on the central new land and transitioned as **EAS North American First Native American sapiens (FNAS)** variants 16 tya. Some of the FNAS variants ventured east over the Cordillean Glacier to the Canadian Rocky Mountains Range at Alberta, Canada. There was a passageway to the south between the Western Cordilleran and Eastern Laurentide Glaciers into and through valleys to below the maximum ice sheet and land in

Northern Montana, USA 12 tya. These FNAS variants were the first known **Eastern Paleo Americans** below the ice sheet maximum on the east side of the Rocky Mountain Range. FNAS variants transitioned as **Native American sapiens (NAS)** variants in the USA without the EEAPS influence 13-10 tya.

European East Asian/pithecanthropus sapiens (EEAPS) variants migrated from northern China and East Russia Beringia into the Alaska-Canadian Northwest Territory, south along the Pacific Ocean ice bridges and the Cordilleran Ice Glacier 17-16 tya. How they migrated along the Pacific Coast to the Americas' in cold climate, negotiating the Alaska-Canadian Northwest Territory and Cordillean Glacier into a new world of North America is a continuing process of fact finding research. They would have followed the sea mammals and fish along Beringia-Alaska-Cordilleran Ice Glacier to the maximum southern glaciation and open land of North America Washington State (USA) west side of the Rocky Mountain Range 15 tya. EEAPS were known as **Western Paleo American**.

The American Native People and Europeans

The First Native American sapiens (FNAS) lived in tribes in Canadian Northwest Territory, and Alaska part along the Atlantic Ocean, now the United States of America (USA). They were hunter gathers in Canada and USA on their land along the Atlantic Ocean and the Great Lakes. Over time FNAS transitioned to Modern Native American sapiens (MNAS).

Norwegians; Norseman, Swedes, and Danes from Europe colonized Greenland in the 900's. They sailed to the southwest in their Danish built longships and settled on an island they called Vinland off the Atlantic coast of North America in the 1000's. They sailed to the east coast of North America; Labrador, islands of Newfoundland and Nova Scotia living in longhouses where their civilization became lost, probably caused partly by close inbreeding and desease.

The Spanish sailed the Atlantic Ocean to the Americas', the islands of Bahamas and Cuba and Florida Territory looking for riches in the 1400's. They migrated to Florida, the Gulf of Mexico, Central America, and the western territory along the Pacific Ocean. The Pagan MNAS aka Indians of Mexico and South America were systematically converted to Christianity belief and faith (Catholic religion spirit god) by Spanish priests. MNAS were slaves and worked the land for the church in return they were given food, shelter, and forced spirit training enforced by the Spanish Conquistadors equipped with guns, swords, rode horses, and did not value other human life, all new to the Modern Native American sapiens (MNAS). They spread white man desease among the MNAS. Also, the rats the Spanish imported in shiped goods, causing desease among the MNAS. The Spanish law required the Conquistadors upon their return, to bring back gold commandeered from the MNAS and advanced ADNAS native population from different tribes. The MNAS waged war with the Spanish in South America.

The Spanish/MNAS variants became Mexicans. The Rapublica de Mexico was established in 1500-1800's. There was war over land the Mexicans claimed land in the southwest territory of the Americas'; Texas, New Mexico, Arizona, and California. In the 1850's the USA government purchased the land from the Rapublica de Mexico territories it claimed; Texas, New Mexico, Arizona, California, north of the Rio Grande River from the Gulf of Mexico, west to El Paso, and west to San Diego at the Pacific Ocean.

The North American New England colonies were established by European settlers and British military on native tribal land of the MNAS populations along the northeastern Atlantic Ocean and the Great Lakes. The British European Military from the British Isle, England, sailed the Atlantic Ocean landing at what is now Massachuetts, to establish new land for Queen Victoria of England in the 1500's. The French were claiming land in eastern Canada and the northern shores of the Great Lakes Territory. The FNAS/MNAS variants established their territory along what are the northern USA states from the eastern flank of the Rocky Mountain Range and eastern Canada. The MNAS lived in communities with different languages. MNAS had confrontations with the British and French over the same territory. The British and French built fortified structures, housed troops, and officers on MNAS land.

The MNAS claimed the land and were at war with the French and the British over possession of their land in 1600's.

Many emigrants from different parts of Europe were sailing to the new world in North America to start a new life due to overpopulation, authoritarian governing, and religious persecution. There was MNAS, European and the forced Africans captured and purchased of human slaves by prominate land owners to work the land in the Americas'.

Plantation owners purchased imported African slaves to work the land, who had no rights 1600-1900. The New England settlers American Revolutionary War for American Independence was for individual freedom of judgement and action from the British rule in the 1770's. These European emigrant settlers established a nation formed for and by their people with their Bill of Rights, Constitution of the people of thirteen united colonies, excluding the MNAS and African slaves.

Great Britain and the USA were at war, 1812-1815.

The MNAS were forced to live on land west of the colonies on what were called land reservations. The USA Government relocated the MNAS to reservations, established trading posts to supplement food, goods, materials, and oxen. The MNAS traded leather, fur materials, pottery, ornamental items, hunting equipment, and other exchange items. The MNAS domesticated the plains wild horses left over from the Spanish Inquisition, killed the buffalo; for food, clothing, shelter materials, and killed other game for food and fur pelts to trade. Some Trading Post managers took advantage of the MNAS for their benefit and profit. MNAS lived in poverty. Later the USA government provided some living subsistance allowcations of limited subsistance at government Provisioning Posts, to be paid with megar government vouchers.

Many MNAS rebelled with vengeance and killing of Trading Posts management, many eastern settlers, on MNAS land the USA Government claimed was open range for settlement. USA broke treaties; not providing enough food and materials to survive when the hunting and gathering was not enough, allowing settlers to live and farm on MNAS land, and USA military force built forts to protect settlers on MNAS land, all in disagreement with the USA government.

Some of the MNAS with initative and goals transitioned to **Advanced Native American sapiens (ADNAS)** and were educated, changed their life style outside the reservation social culture, a new language, and exposure to a technical world. Later it included schooling on the reservation in the American way. However, it has not provided jobs for those on the reservation who live on government entitlement.

The industry developed north, the Union vs the less developed south Confederacy of the USA, entered into the American Civil War between the States over civil rights in the 1861-1865. United States entered two World Wars 1917, 1940, and many foreign conflicts of different thinking and actions. We all lost loved ones to violence and behavior of others in disagreement.

Anger toward others is a learned emotional stress and behavior manafested by violence: civil unrest, negative attitude, mental disorder, stealing by taking others property or personal identity or mind thinking, and misuse of drugs and guns. We need to learn how to communicate properly, understand the other side, and learn to love one another. Living in a destitute environment is a source for emotional stress, negative behavior, and violence. Positive thinking and behavior action education does not come easy. There is a need for individual positive incentive to break loose from the negative attitude and depression of a living environment before it destroys civilization. It is possible in the computerized digital data world to provide artificial intelegence and robotic machines. But, until we event machine reasoning ability Homo sapiens will thrive from experience, knowledge, skills, and abilities.

SUMMARY

And in the begining events evolved from a single state element of gas matter when; mass, space, time, and energy all came togather and exploded. The non-living thing evolved as many universes by Natural Selection.

Part 1 is devoted to Relation Theory and Natural Selection. This study explores the non-life origin events of the universes that evolved 14 billion years ago. The solar system matter is responsible for living life, family of the Primate Hominids, human Homo sapiens evolution and ancestry. Details about the events are explained in Section 1, 2, and 3.

Sections are detailed with evolved Origin, Life, and Family. Non-Life Origin evolved when expanding universes were created, new gas matter was developing and our plasma matter sun was forming. The violent volcanic planet Earth was formed cooling into a solid matter mantle and water. Life and family is possible because of organic micro molecule events of Amino and Deoxyriboneuleic Acid protoplasm, necessary for all life by Natural Selection.

Non-Life Origin of Matter: early universes evolved from a hydrogen and helium gas matter explosion of the, "Thing", from a single point in space. The events of origin from nucleosynthesis, over time, evolved from elements of gas atoms. Galaxies and solar systems formed. In our solar system heavy elements condensed into liquid matter. That matter expanded and the resultant explosion formed our proto sun nebula. Later as time passed the gas matter, gravity and temperature rose, the hydrogen fused to form helium, and exploded into a full star as Earth's plasma matter sun. Other planets took their place in orbit around the sun according to their specific gravity. Plasma matter planet Earth was cooling into a solid matter rock mantle in stratum layers, with a liquid matter core. The orbiting debris within Earth's gravity in conjunction with our sun's plasma solar winds formed our moon by accretion. Earth had no atmosphere and was bombarded by meteorites and solar winds vaporized the oceans releasing hydrogen and helium into space.

Earth was a violent volcanic erupting planet, forming new solid matter (rock stratum), emitting anoxic chemicals into the seas of water. The atmosphere was void of ozone and ultraviolet radiation bombarded the Earth. The cooling of Earth caused great rifts in the oceans. Tectonic plate movement repositioned great land masses. Some rift plates collided with such great force making new solid matter, forming mountains, where the land pushed up over the solid matter.

Earth system events evolved; Snowball Glaciation, Paleo-Magnetic Era, K-T extinction, time, and climate.

Origin of Life: time passed, stratum element matter formed amino acid Ribonucleic Acid (RNA) combind with protoplasm Deoxyribonuclic Acid (DNA) by Natural Selection essential for all living matter. Life in the form of single cell organic micro-molecules organelles formed from stratum RNA and DNA evolved in anoxic sea water. Microlife in a volcanic and anoxic environment had challenges establishing and surviving the ultraviolet radiation, carbon dioxide, and anoxic sea water in the Old World nascent life. Early single cells produced a waste product Oxygen. It excaped into Earth's space to form Earth's atmospheric ozone boundary. Problem was these microorganelles could not reproduce. Ultra violet radiation exposure caused some micro life to cease to exist due to not adapting to its environment.

Life system events: As time passed, a new organic micro multi-cell organism development in a membrane of living cells evolved from the compounds of organic element matter combination contained in the DNA of each cell in the double helical genome of life that could reproduce. Survival of the fittest cells prevailed and were able to grow in size. The developing changes in form of the strongest adapted to its environment. There were divergening genus and species that transformed and transitioned as the variants from changes in mutant traits to the genome DNA markers transmitted to each successive next generation.

Gene mutations resulted in a new and different form by a process of natural selection. The surviving gene cells changed, growing, and adapting to new environments. The experiences over the life of a form influenced small minute differences in the DNA passed on to the next generation by two different compatible forms sexual exchange.

Organic micro cells transitioned to the next generations of life as Micro-cellular organisms. Ancient creatures grew larger in the oceans. Primates Placental Mammal animals appeared on land. Primate Common ape diverged from the monkey. All evolved in the Old World Pongidae before 20 mya.

See population and location charts identifing species living in the referenced text time as defined, in Africa, Europe, Asia, Australia, and the Americas'.

Section 2 opens with the Hominidae great apes. It was the most challenging of the research developing the phylogeny for populations and species of Old World Pongidae, and New World Hominidae Primate; Common ape, Great Ape, hominid Australopithecus, and Homo.

Origin of Family: The diverged group of early specialized Australopithecus hominids adapted to their environment. In the New World many species and variants evolved and passed on gene trait differences linked to the development of Hominidae Primate Apes, Primate Great Apes, and a specialized australopithecus hominid ape; diverged, transformed, and transitioned to Homo apeman, manape, and man as Homo sapiens.

Hominids lived in trees until forced to live on the ground environment due to climate changes. They learned to adapt to survive. The change in diet included scavanging meat kills and grubbing for tubers beneath the ground. Most were nomadic and frequently migrated in their way of life. Making small animal kills was practiced because it made it more palatable to eat.

Homo apeman, manape, and man; learned to hunt using Stone Age Technology pebble to sharp flaked to butcher animals. They advanced to using spears with stone points they fabricated. They learned to live togather in social culture, civilization, use fire, and trade. Populations grew and food resources declined. It became difficult to survive. There was aggression, violant behavior, and cannibalization of their kind to survive.

The old world and new world evolution of ancestors is still being discovered. Family ancestors linked to life are; organic micro-molecules, micro organisms, ancient creatures, placental mammal animals, primate monkey-ape, and the divergence when monkeys diverged from apes. Some Common Apes transformed and evolved as Great Apes. The specialized australopithecus ape are the ancestor forms we share with variations of apeman, manape, and man. These discoveries have many scientific unknowns in evolutionary time. We know more about Australopithecus and Homo genus development over time. On the transformation trait advancement toward Homo (human) genus there were genome mutants in Australopithecus ancestry in physical body form discovered in fossils with advanced trait features leading to mans evolution.

Apes were walking up right in the trees with stiff backs and feet with a spayed big toe. When they moved to the ground over time they evolved a back with a curved spine. The pelvis changed and become the center of gravity for bipedal upright walking. The paralleling of the big toe with the other toes provided greater forward push off when walking. They grew in size thriving in their adaptation to their environment.

Chart 1 Phylogeny Old World Pongidae Common Ape New World Hominidae Great Ape 124-4.5 mya, includes population and species living at those times.

The Homo was determined from brain size. Some Primates evolved as ancient, early, late, advanced, and anatomic transformations in physical form. They had bigger brains and were classified as Homo genus. There were those who did not adapt to their environment and ceased to exist. Apeman and manape were Homo. As the species variations evolved there were many different forms. One of the later last human transformations was Homo neanderthalensis (NEA). It was all man, cousin to Homo sapiens. Only Homo sapiens have survived as the fittest by natural selection. Todays Homo sapiens are a mix of many cultural variants throughout the civilized world.

The apes and homo's developed stone tool Olduvai Technology to assist them in adapting to their environment of everyday survival. These scavengers used the first primitive tools of stone core hand ax for meat carving, cutting and crushing bones. The risk in the hunt advanced to direct contact capture of small animals. They used wooden thrusting spears sharpened on one end for killing the small animals. The other end had a rounded knob to assist in the thrusting action of the kill and for grubbing in the dirt for tuber growth.

Later, as hunters they learned by experience, developed advanced stone tool Acheulean Technology. They applied their knowledge learned to the skill of knapping stone core shaped into blades, axes, hammers, and throwing weighted stone flake spears used in hunting and butchering of larger animals. Improvements skills were made to worn tools by sharpening to sharp edges.

The meat in their diets was responsible for Homo brain development, neurological mutations, changing their ability and behavior to think and reason with the advanced development of the brain frontal lobe. They were marking objects indicating their abstract thinking and symbolic behavior to plan and innovate.

Social cultural behavior concepts and community living (civilization) were starting to emerge with the development of the brain frontal lobe responsible for emotion and compassion. Modern thinking transition introduced behavioral thinking, cognitive traits, and creative thinking in a new way to support more sophisticated social skill and tool technology needed to survive. The symbolic thinking increased with the ability to abstract, analyze the past, and anticipate the future. Improved strategic thinking by reasoning allowed them to plan, create art, use a fully articulated speech, and a language to pass on information more efficiently.

Civilization within a social culture was living together for the common good of all. Agriculture became the common way to survive. Each harvested goods from the land and domesticated cattle, goats, and reindeer. They shared, established trade, and commerce of their bounty with others. This was a major social cultural change to thrive. There were those who exploited the common population and their bounty in the form of authoritiarian governed controlled power over the population to conform to rules or be punished. Those mafias contributed nothing to the social culture.

The Elite Homo sapiens were evolved variants of arachaic H. sapiens (AS), transitioned to early H. sapiens (ES), transitioned to early modern H. sapiens (EMS), transitioned to late modern H. sapiens (LMS), transitioned to advanced anatomic modern H. sapiens (AMS), and transitioned to

Homo sapiens of today (MAN). The NEA was a cousin of modern sapiens transitioned from H. antecessor variant beget from a questioned HEB or ERG variant. NEA lived a brutal life and became extinct.

There was a conflict concerning the subjective standard of identification of Archaic H. heidelbergensis (AHEB) variant having H. ergaster (ERG) traits and its origin alleged to have been from African, African H. Rhodesiensis (RHO) beget by ergaster/erectus variants may have been incorrectly identified from a thick skull discovery in European Russia. Evidence suggests RHO did not migrate into northern Europe. Determination of HEB was from a thick jaw skull cap discovery in stratum dated to 600 tya from Germany. It was used as the comparison basis reference throughout Europe to Spain to identify HEB claimed thick skull human fossils.

There were great migrations from Africa into Europe, Asia, and the Americas'. There are theories of migration into the Americas': European LMS in Asia transitioned to the European Asian sapiens (EAS) and migrated over Beringia, Eastern Russia, into south central Alaska Northwest Territory. EAS transitioned to the North American First Native American sapiens (FNAS). Some migrated over the western Cordilleran Glacier and between the eastern Laurentide Glacier passage way, through the Canadian Rocky Mountains Range (RMR) at Alberta, Canada, and south through the valleys, following the herd animal trails to the glacier maximum into what is northern Montana, USA. On the east side of the RMR, FNAS variants migrated east, known as Eastern Palo Americans. FNAS were not gene influenced by European Asian sapiens East Asian Pithecanthropus sapiens (EEAPS). FNAS transitioned as the Native American sapiens (NAS), east along the Laurentide Glacier maximum land to the Atlantic Ocean. NAS transitioned to Modern NAS (MNAS). Groups of MNAS migrated south to Florida, around the Gulf of Mexico coast, south into Central America Mexico, the northeastern parts of South America, and to the Atlantic coast in Brazil. MNAS transitioned as todays Advanced NAS (ANAS).

The isolated EEAPS theory: migrated from south Eastern Russia and north China, along the Pacific Ocean coast, on the west side of the Rocky Mountains, into Washington State, USA, and the Pacific Ocean coast. They were the Western Palo Americans.

Homo sapiens are the survivors of transformation, transitioning of species and variants, who survived as the most fit.

Section 3 expands on the technology aspects of some of the text data, testing methods, ethics (when there is no recourse form responsible truth and accountability), dating identification techniques, climate, and life events.

Innovation is responsible for developing change. Change improves on what is established by standards. Standards are used to meet and exceed excellence by performing tasks in a responsible ethical and moral manner with the pride of being accountable. Honor, trust, and respect are individual achievements. Rewarding for excellent performance and accomplishments above the established standards and rules should be common practice.

Testing, analysis equipment, and other technology advances are applied and used to perform research, analyze age, and identification of fossils and stratum determinates are within limits. Tool dating uses a verity of technologically innovative equipment in determining age within limits. With the development of genetic analysis and testing from DNA sampling by a technician can compare life relationships unique to human ancestry.

Part 2 African European violence and behavior differences, appearances of class, and their catagories were changing. They killed each other to survive thousand of years ago in Africa. In Europe there were those who pilfered the land and the population. Some with their ill behavior, defensive attitude, power and control over populations caused violence and warfare. Many Europeans with their mixed ancestry, departed from their home country, and risked a voyage across the Atlantic Ocean to the Americas' to start a new life. The MNAS and ADNAS fought foreign land and population pilfers. ADNAS today are forced to settle on land on reservations with government subsistence. The USA government forced the tribes to relocate to reservations. Sounds like a time passed, when there were European Vikings and Teutons, however, this pilage has a modern approach.

In our hyper-modern technical world we deal with artificial intelegence and robotics making Homo sapiens tasks more reliable, with better accuracy, and production. In the bio-technic world we deal with the very essence of life. RNA and DNA genetics can explore our heredity and variation of Homo sapiens.

Because we live in an imperfect world, we co-exist with imperfect Homo sapiens with different views about everything. It takes all kinds of individuals with different brain capabilities and thinking, the good, others bad, and the mental and criminal evil. Such is life we live in. We need rehibilition to be more effective.

In conclusion, please listen to others and understand what was said before you communicate an agreement and action plan or statement. Each side should result in a win-win agreement. This approach to conflict is intellectual emotional communication to reduce stress and anger. We earn trust and honor. There is no recourse from the truth.

This scientific history study and research is the result of over 50 years about Homo sapiens origin, life, and family ancestry having many unanswered scientific issues associated with facts, incorrect findings, wrong identification, opinions, and theories about who we are and where we come from. The author has a Family Tree ancestry and decendence with over 600 individual Homo sapiens.

www.ingramcontent.com/pod-product-compliance
Lightning Source LLC
Chambersburg PA
CBHW030742180526
45163CB00003B/887